Havoc and Reform

HAGLEY LIBRARY STUDIES IN BUSINESS, TECHNOLOGY, AND POLITICS

Richard R. John, Series Editor

Havoc and Reform

Workplace Disasters in Modern America

James P. Kraft

JOHNS HOPKINS UNIVERSITY PRESS   BALTIMORE

Johns Hopkins University Press
2715 North Charles Street
Baltimore, Maryland 21218-4363
www.press.jhu.edu

Library of Congress Cataloging-in-Publication Data

Names: Kraft, James P., author.
Title: Havoc and reform : workplace disasters in modern America / James P. Kraft.
Description: Baltimore : Johns Hopkins University Press, 2021. | Series: Hagley library
    studies in business, technology, and politics | Includes bibliographical references
    and index.
Identifiers: LCCN 2020026094 | ISBN 9781421440576 (hardcover) | ISBN
    9781421440583 (ebook)
Subjects: LCSH: Industries—West (U.S.)—History—20th century. | Disasters—West
    (U.S.)—History—20th century. | Work environment—West (U.S.)—History—20th
    century.
Classification: LCC HC107.A165 K73 2021 | DDC 363.110979/09045—dc23
LC record available at https://lccn.loc.gov/2020026094

A catalog record for this book is available from the British Library.

Special discounts are available for bulk purchases of this book. For more information, please contact Special Sales at specialsales@jh.edu.

To Ed,
for all the good advice

Contents

Acknowledgments   ix

Introduction   1

1   Disasters at Work: A Brief Background   16

2   Chemical Plant Explodes in Texas City!   43

3   Airliners Collide over the Grand Canyon!   74

4   Hospitals Collapse in Southern California!   104

5   MGM Grand Burns in Las Vegas!   135

6   Federal Building Bombed in Oklahoma City!   165

Conclusion   195

Notes   209
Essay on Primary Sources   245
Index   249

Acknowledgments

Historians often work in utter isolation, contending with the many challenges of research and writing. But bringing our projects before readers is never done singlehandedly, or even with the help of a few extra hands. It took an unusually large crew of individuals and institutions to complete this project, and I have long looked forward to acknowledging this assistance.

During the project's earliest stages, when its boundaries were still unclear, I received much-needed support from my home institution, the University of Hawai'i at Mānoa. The staff at the Hamilton Library gathered an assortment of old newspapers, government documents, and other resources that helped to map out this project. The assistance provided by librarians Gwen Sinclair and Sheila Ayson proved particularly important. The History Department and the College of Arts and Humanities contributed to the project by offering occasional travel grants for research as well as opportunities to step away from my teaching responsibilities. Submitting funding requests and planning research trips almost always involved the department's administrative support specialist, Sue Carlson, who could not have been more obliging.

Moving the project beyond shallow waters required guidance from research specialists on the "mainland." Initially, this involved traveling to the great state of Texas. At the National Archives in Fort Worth, I received excellent direction from Michael Wright and further assistance from Rodney Krajca, Barbara Rust, and Ketina Taylor. I received more support from Norie Guthrie at Rice University's Woodson Research Center in Houston and Jennifer Nuzzo, Elisabeth Sargent, and Jennifer Sessa at the Houston Public Library's Metropolitan Research Center. Rebecca Snow and Beth Steiner pointed me to a wealth of resources at the Moore Memorial Public Library in Texas City, and Sarita Oertling and Julie Trumble provided assistance at the University of Texas's Moody Medical Library in Galveston.

Librarians and archivists in California and Nevada were also generous with their time and expertise. At the Richard M. Nixon Presidential Library and Museum in Yorba Linda, Carla Brasswell and Gregory Cumming steered

me toward useful archival collections, as did their counterpart Kelly Barton at the Ronald Reagan Library and Museum in Simi Valley. At the Huntington Library's Munger Research Center in Pasadena, staff members Morex Arai, Stephanie Arias, Sara Couch, and Samuel Wylie were very accommodating. Ross Kendall and David Sigler led me to valuable materials at the California State Northridge Library, and Rebecca Kohn and Diane Malmstrom provided access to electronic resources at San Jose State University. I received assistance from Peter Michel and Joyce Marshall Moore in the Lied Library at the University of Nevada, Las Vegas, and Teresa Wilt at the Nevada Legislative Council Bureau in Carson City. Staff members at the Nevada Historical Society in Reno not only alerted me to materials germane to my research but also encouraged me to publish "Disaster in the Workplace: The MGM Grand Hotel Fire of 1980," *Nevada State Historical Quarterly* 56, no. 3–4 (Fall–Winter 2013): 167–86. The journal's former editor-in-chief, Michael Green, guided that article through the review process, and Mella Harmon, Sheryln Hayes-Zorn, and Hillary Velazquez brought it into print.

Research specialists in Oklahoma and Missouri also played vital roles in this project. I am especially beholden to the staff at the Oklahoma City National Memorial and Museum. The manager of the museum's collections, Helen Stiefmiller, went beyond the call of duty to support this project, and Joanne Riley and Lynne Roller made sure no side streams were left unexplored. At the University of Oklahoma in nearby Norman, Special Projects Librarian Karen Antell helped me explore a bundle of local newspaper collections. I found the staff at the State Historical Society of Missouri in Kansas City accommodating, too, especially David Boutros and Whitney Heinzmann.

I have long considered oral history to be a useful research tool and often relied on the recollections of disaster victims to search for the larger meaning of these tragic events. I interviewed several people who were employed at the MGM Grand when it caught fire, including Patty Jo Allsbrook, Cathy Beane, Jim Bonaventure, Eileen Daleness, Kathy Griggs, and Sherry McClaren Walberg. I also interviewed people who were working in the Alfred Murrah Building when it collapsed in 1995; namely, Janet Beck, Dianne Dooley, Dot Hill, Germaine Johnston, Dennis Purifoy, Beverly Rankin, Florence Rogers, Jennifer Takagi, Susan Walton, and Richard Williams. Looking back on personal tragedies is often difficult, and I am deeply appreciative of the contributions that each of these individuals made to this project. My

travels included a side trip to Arizona's magnificent Grand Canyon National Park, and I would like to thank park ranger Daniel Pawlak for identifying the crash sites of the 1956 air tragedy. Anyone who has seen the stunning gorge knows it is simply indescribable, and those particular sites are breathtaking.

If a picture is worth a thousand words, as the saying goes, then I owe a huge debt to the individuals and institutions that helped me collect illustrations for this book. I began looking for potential illustrations shortly after launching this project, and a few of the images reproduced herein came from specialists previously mentioned. But the search involved many other generous people, including Simon Elliot and Molly Haigh at the University of California's Charles E. Young Research Library in Los Angeles; Melody Bragg at the National Mine Health and Safety Administration; Laura Saviano at Ross Barney Architects; and Joel Draut, Barbara Estrada, Elizabeth Mayer, and Matt Richardson at the Houston Public Library. Equally helpful were Jeric Griffin at Trinity Mother Frances Health System, Becky Tyner at the New London School Museum, Cynthia Summers at Northern Arizona University, Carol Butler of Brown Brothers, Matthew Lutts at the Associated Press, Karine Mnatsakanian at Getty Images, Brian Humphrey and Erik Scott at the Los Angeles Fire Department, Jeff Suess of the *Cincinnati Enquirer*, Lily Berkhimer at the Ohio History Connection, and Cheri Daniels at the Kentucky Historical Society.

I owe the greatest debt to the individuals who read and commented on all or parts of this work. I received some early inspiration from a former colleague at the University of Hawai'i, the late Idus Newby, whose wise counsel served me well for many years. Once I had a full draft of the manuscript, I sent it to my former doctoral advisor at the University of Southern California, the late Edwin J. Perkins (to whom I have dedicated this book). As always, Ed found ways to broaden my historical perspective while simultaneously encouraging me to follow my own instincts and interests. My friend William Childs, with whom I worked at the Ohio State University, read a revised version of the manuscript to my advantage, and Rebecca Clifford, former journals editor at the University of Hawai'i Press, carefully proofread the work. My cousin Kay Exum Fowler, librarian extraordinaire, read and polished it, too.

My association with Johns Hopkins University Press dates back to the previous millennium, and it is a privilege and honor to be associated with

the press and its staff. Robert J. Brugger encouraged me to pursue this project long ago, and his frank comments on the early draft sharpened my sense of direction. Gregory Britton and Laura Davulis welcomed my initial inquiries about publishing the work, and William Krause and Matthew McAdam skillfully guided it through the review process. Richard John, the editor of the fine series in which this book appears, expressed an interest in this project as soon as it came to his attention, and I have benefited immensely from his multiple readings of the work. I also benefited from the advice of three anonymous reviewers, each of whom opened my eyes to new possibilities. Ashleigh McKown meticulously copyedited the final draft, while Juliana McCarthy, Hilary Jacqmin, Kyle Kretzer, and Kathryn Marguy produced and promoted it.

Before striking the sails on this project, I wish to acknowledge the many kindred spirits whose brief words of encouragement kept this project afloat. In 2009, I presented a paper on disasters and workplace safety at the International Science in Society Conference, held at the University of Cambridge, and received inspiring feedback from the audience and co-panelists. I discussed the subject again at the 2013 National Social Science Association in Las Vegas and found more reasons to stay the course. Meanwhile, I began to discuss this subject with friends and fellow historians, and their questions and remarks often brought distant horizons into sharper focus. I send my warmest aloha to longtime friends Sheldon Anderson, Mansel Blackford, David Hanlon, Peter Hoffenberg, Louise McReynolds, André Millard, and Arthur Verge, who all listened patiently whenever I mentioned this work. I also received encouraging words and messages from several other colleagues and cohorts, including Carl Abbott, Joel Bradshaw, Roger Horowitz, Matthew Lauzon, Mark McNally, Steve Ross, Philip Scranton, Yuma Totani, and Theodore Jun Yoo. Lastly, I want to acknowledge the enduring, positive influence of my brother, Robert A. Kraft, to whom I still look for guidance. I also wish to thank my amazing wife, Renée Arnold, for once again helping to navigate the occasional storms. It's been an adventure as well as a journey.

Havoc and Reform

Introduction

Workplace disasters have upended the lives of millions of Americans. They have resulted in deaths and disabilities as well as the loss of jobs, properties, and savings. The worst of these tragedies—those that left the largest historical footprints—have been more than just destructive. Some have overwhelmed the resources of private charities and public agencies, even in recent times when disasters have generally warranted assistance from the federal government. Some have also inspired safety reform movements that reshaped work environments, in ways that partially compensated for the deaths, suffering, and displacements that such tragedies caused.

These entangled patterns of havoc and reform drew my attention more than a decade ago, as I was completing my previous study of labor relations in the Las Vegas resort industry.[1] Reading news clippings about historic strikes in the industry, I came across a few unforgettable photographs of the 1980 MGM Grand Hotel fire, a calamity that killed eighty-five people, including nine employees. Staring at pictures of thick black smoke hovering over the huge resort, I had to presume the fire must have displaced the entirety of the giant facility's workforce. What happened to those employees? Did they take jobs elsewhere? Did some leave Las Vegas, and if so, under what circumstances and in what frame of mind? And what happened to the MGM itself? How badly did the fire damage the resort, and how did its chief officers deal with the crisis thrust upon them? I also wondered about the safety of other large resorts in Las Vegas. Had there been subsequent hotel fires in the city, and if not, why not?

Thinking about such questions, I began to see a neglected area of scholarly inquiry. Historians have not ignored disasters that wrecked large work-

places, but our accounts of these tragedies are often fragmented and incomplete. We rarely consider what these events had in common or how they differed. We also tend to sidestep questions about the settings in which these events occurred and the people whose lives they claimed. Moreover, we have not often asked about the individuals who had to answer for such deaths and injuries, or how these calamities triggered new safety regulations. Time after time, these events have forced the nation to rethink old ways of design and building, and in doing so encouraged changes that registered as "progress." A cursory look at textbooks in the closely related fields of business and labor history sheds light on the problem. These books analyze a constellation of challenges that once confronted business enterprises and their employees, but they say almost nothing about disasters like the MGM Grand fire.[2]

Much of what we know about these events comes from journalists and popular writers whose articles and books are often gripping but not always enlightening. The literature helps us understand critical details about great tragedies, but it seldom speaks to commonly asked questions in American business and labor history, such as how unexpected crises affected the structure and operation of firms, or how they affected the daily routines and attitudes of their employees. Nor does the literature typically explore the relationship between disasters and public policy. It often suggests that industrial accidents had beneficial outcomes without clearly explaining how or why, leaving the impression that any positive reforms that these tragedies inspired simply fell from the sky, like heavenly rain from passing clouds. Who championed these reforms, and how were the reforms implemented?[3]

Scholarly or journalistic, the bulk of this literature focuses on a few ghastly accidents that occurred around the turn of the twentieth century, somewhere in the nation's northeastern states. The focus is understandable: this was an age when large corporations began to control the process of industrialization and often pursued strategies at the expense of workers' health and safety. Millions of new immigrants joined the nation's labor force, including unprecedented numbers of women and children. The concentration of capital along with steady population growth transformed sleepy towns across New England into bustling nerve centers of production and simultaneously created large coal-mining communities throughout the hills and plateaus of the Appalachian Mountains. Jobs in these areas of the economy were exceedingly dangerous, and employers seemed mostly unconcerned

with the risks taken by their employees. Despite the fiery protestations of labor leaders and social activists like "Big Bill" Haywood and Mary "Mother" Jones, occupational fatality rates soared. By 1941, when the United States entered World War II, industrial accidents often killed dozens of workers in one fell swoop.

Though the nation's work environments are far safer today than they were more than a century ago, their sudden and unexpected destruction remains a fundamental part of our advanced, interdependent society. The persistence of these events raises its own questions: Where have these disasters occurred, and what sorts of structures did they damage? Whom exactly did they affect, and how? What, if any, safety reforms did they spur, and how exactly did such reforms materialize? Did business leaders and political elites actively support safety proposals, and if so, again, why? Further, what do these events tell us about the modern world in which we live? Might they help us contemplate the paradoxical features of modern industrial and technical advances?

Addressing such questions is no simple task. It requires the study of contrasting industries and public agencies as well as different occupational groups and work settings. It also demands an appreciation of local and state building codes, federal safety legislation, and new safety technologies. But challenging questions underpin all historical studies, and after reflecting on a few of my own, I began to envision a manuscript about a most compelling subject: *modern-day disasters that wrecked large places of employment far from the nation's early industrial centers.* I grew increasingly interested in the *consequences of modern disasters, not just for working people but also for work sites and public policy.* A book about these matters, I anticipated, would likely challenge a few common assumptions about calamities in the realm of work.

This book has multiple themes and subthemes, but it delivers a singular message: workplace disasters are not merely phenomena from the distant past or rare occurrences that affect only a narrow range of occupational groups. They have never fit neatly into any single category or followed any standard script. Even in the early stages of the Industrial Revolution, when textile factories dotted the eastern landscape, these events occurred in all sectors of the American economy and in different sections of the nation. Their victims were young and old, male and female, black and white. Many of the victims were recent immigrants new to the workforce, but a number of

others were native-born citizens who had been there for years. In addition—
and this is a crucial point—many were not even employed in the affected
workplaces. In the transportation and service sectors of economic life, disasters in one's place of employment often killed and injured more customers and patrons than employees, sometimes by ratios of more than ten to
one. Any efforts to minimize hazards in those settings therefore had implications for public safety as well as the safety of workers.

These stark realities hold special importance in twenty-first-century
America, where service-oriented industries produce more jobs and wealth
than factories and mines. Modern structures like airliners, hospitals, and
hotels may not look like work sites to most of the people within them, but
they nonetheless employ sizeable labor forces. When the MGM Grand Hotel
caught fire in 1980, the posh resort employed some forty-five hundred people. The place was open twenty-four hours a day, seven days a week. Its
employees were certainly not a homogeneous group or people without personal lives and career goals. Resort workers filled all sorts of job positions,
including those that required special skills and talents and were thus highly
compensated. Like their counterparts in work environments of earlier times,
they had families and loved ones as well as concerns about issues like housing costs and health care. But so, too, did the hundreds of guests in the resort's hotel rooms, casino, restaurants, shops, and bars. To put it all too
plainly, when modern-day workplaces crash and burn, many different worlds
turn upside down. All sorts of towers topple.

Making work sites safe has proven to be a challenging and time-consum-
ing process. It has involved a multitude of individuals and institutions at all
levels of government, and it has required compromise between competing
interest groups and different political parties. Defenders of "state rights" and
even "city rights" have further complicated the process, as have officials who
demanded small government at any cost. But when disasters affected large,
eclectic groups of people, safety reform tended to move ahead more quickly.
Longtime anti-regulatory conservatives started speaking like progressives,
and legislators at both the state and federal levels of government listened
carefully.

Unlike my previous books, this work does not rotate around the axis of
capital versus labor. It is less about capitalist imperatives than the unintended consequences of industrial and technological change. In other words,
it is not about how societies divvy up the capitalist pie, or whether certain

occupational groups deserve a larger piece. Its focus is primarily on hazards of the modern world and how they have altered lives. But its larger concern is with the question of how the nation has tried to minimize the threats that these hazards pose to life and property, and with what success. The work does not attempt to measure the number of accidents or fatalities that safety reforms have prevented, however, or to venture into other areas of counterfactual history. Some consequences of calamities are simply beyond our ability to comprehend.

What ties this work together is the conviction that rapid industrial and technological advances form a sharp, double-edged sword. This argument is certainly not new—historians and social theorists have made it for generations. Among the first Americans to do so was the political economist Henry George, whose writings enthused many social activists and reformers of his time. In *Progress and Poverty* (1879), which investigated the roots of social and economic inequality, George brought the two-sided character of material growth into sharp focus. "The march of invention has clothed mankind with powers of which a century ago the boldest imagination could not have dreamed," he wrote. "But in factories where labor-saving machinery has reached its most wonderful development, little children are at work."[4] George reiterated the point in *Social Problems* (1886): "It is clear that the inventions and discoveries which during this century have so enormously increased the power of producing wealth have not proved an unmixed good. Their benefits are not merely unequally distributed, but they are bringing about absolutely injurious effects."[5] Better than most of his contemporaries, George understood that industrial society had produced a cauldron of hazards and risks.

This line of reasoning runs forcefully through the new and still emerging field of risk theory, on which part of this work stands.[6] Two sociologists who helped to found the field—Anthony Giddens and Ulrich Beck—provided some of the tools I needed to complete this project. In *The Consequences of Modernity* (1990), *Risk Society* (1992), and more recent writings, these scholars warned that the productive forces of modern industry and technology have unleashed threats to public safety on a massive scale. Advanced nations like the United States, they pointed out, are allocating more and more of their resources toward administering and monitoring these threats, and in the process establishing new risk-managing structures. The ever-increasing production of risks, they added, has forced public leaders to make progressively more pledges of security, which in turn encouraged greater regulation

of economic life. As Beck explained in *Risk Society*, "The promise of security grows with the risks and destruction and must be reaffirmed over and over again to an alert and critical public through cosmetic or real interventions in the techno-economic development."[7] Anyone living through the corona-virus pandemic of 2020 can see that risk assessment and management has moved to the foreground of public affairs. In this book, I have explored two interrelated questions of concern to risk theorists: *how modern risks have been produced* and *how societies have assessed and managed them.*

This book has a relatively simple organizational framework. Its opening chapter explores workplace tragedies and safety reforms in the century before World War II, when the United States passed through the main stages of the Industrial Revolution to become the world's leading producer of agricultural and manufactured goods. Though valuable on its own, this opening chapter basically serves as a prelude for what follows it: five case studies of postwar disasters and their impact on people, places, and policies. Stitched together, these studies help us to see how modern industrial and technological developments created new risks for American workers and everyone around them. They also provide a sense of how the nation has confronted hazards and challenges of the modern world.

Motivated by the MGM fire, I kept my focus on disasters in the American Southwest. Even broadly defined to include Southern California and all of Texas and Oklahoma, this vast region of the country was still sparsely populated and relatively undeveloped before World War II. The region matured rapidly after the war began, however, and more than a few tragic events accompanied its maturation. To give more depth to this story, I have selected disasters that occurred in different areas of the region, different sectors of the economy, and different decades of the half century after the war. Presented chronologically, these studies offer a unique and sobering outlook on the rise of a now vital, integral part of the national economy. They also underscore the ubiquity and persistence of workplace disasters in American history. The greater Southwest—or the *Western Sunbelt* it may best be called—has not often been associated with workplace tragedies of any kind.[8]

Historians have had much to say about World War II and the American West, and this work builds on their ideas. One of the founders of modern western history, Gerald D. Nash, said long ago that the war "transformed" the West. As Nash pointed out, only 12 percent of the American people lived

west of the Mississippi River in 1941, when the United States entered the war. The region, Nash concluded, was therefore less "rigid" than other parts of the nation and more susceptible to change. To Nash, the prewar West was essentially a colony of the East, a place whose economy depended on an inflow of eastern capital and the export of raw materials and agricultural crops. The war jolted this paradigm. Between 1941 and 1945, the federal government spent a staggering $70 billion in western states (or almost $1 trillion in 2020) to finance defense-related projects. The money helped to establish new shipyards, aircraft factories, steel plants, and many other business enterprises that lured millions of depression-weary, job-hungry Americans westward. According to Nash, these developments laid the foundations for decades of regional growth and thus "liberated" the West from its eastern masters.[9]

Nash's thesis conjures images of hardworking pioneers taming the land, turning what was once wild into something more civilized. As the historian himself recognized, however, the war's impact in the West was much more ambiguous than this scenario implies. Wartime spending in the region went primarily into areas where mild climates favored nonstop production. The region's southern tier of cities and states therefore changed most significantly as a result of the war. These changes, of course, were not necessarily beneficial. In communities where federal spending was heaviest, the war put an immense strain on public services. In some places, even clean water was in short supply.[10] But these problems were mostly local issues as opposed to national concerns. They did not seriously threaten the health and safety of citizens. This book suggests that wartime developments in the region created the conditions for major calamities there.

The worst wartime tragedy on the home front best illustrates the point. On July 17, 1944, back-to-back explosions ripped through a newly expanded naval munitions depot along the eastern margins of San Francisco Bay, in the waterfront community known as Port Chicago. At the time, the naval facility was a vital docking area for warships deployed in the Pacific. The ships dropped anchors there to acquire bullets, torpedoes, and other explosives that came by rail from a Nevada armaments factory. When the depot blew up, for reasons still unknown, more than 320 people died, most of whom were young black servicemen who handled the depot's dangerous cargo. As historians have explained, this disaster revealed blatantly discriminatory policies in the navy, and by implication across the country. It loomed large

in the decision to desegregate the armed forces in the early postwar years and encouraged the postwar civil rights movement. But the tragedy also laid bare a simple truth—what went up in the fast-changing West sometimes came crashing down, and with grave consequences.[11]

The ironies and contradictions of the war's impact in the West remain murky, mysterious, and mostly unknown.[12] Among the historians interested in this subject, Roger Lotchin stands out. In his studies of twentieth-century California, Lotchin challenged Nash's argument that the war transformed the West. Before the war began, he noted, California had grown accustomed to sudden surges in its population and to rapid economic growth. The war, Lotchin stressed, merely accelerated trends and processes long in the making. But the historian went a step further with his conclusion, saying that the war "set the stage" for California's future problems. By spiking the state's population, for example, the conflagration created more traffic congestion, pollution, and housing shortages. It also hastened the rise of ghettos and slums, in turn heightening racial tensions. "In some respects," as Lotchin put it, "the effect of the war transformed California cities toward the future; in others, the conflict created disruption and regression." This gritty underside of wartime developments represented what the scholar called a "despondent corollary" to the Nash thesis. This corollary weaves its way through each of my case studies.[13]

To be clear, I am not saying that World War II singlehandedly reshaped the landscape of the West or solely determined social or economic patterns in the region's Western Sunbelt. Historians have made it abundantly clear that multiple factors account for the growth of the modern West, including its southernmost cities and states. In a recent study of Phoenix, Arizona, for example, Elizabeth Shermer argued persuasively that a cadre of well-connected lawyers, bankers, retailers, and legislators transformed Sunbelt cities into postindustrial centers of trade and finance. Working largely through chambers of commerce and city councils, these groups lured corporations to the West with promises of tax breaks and minimal government regulations. The larger point is that wartime spending did not fully explain the rise of the greater Southwest. Postwar "boosterism" also freed the region from its prewar status.[14] But these historical forces were not mutually exclusive. Indeed, they dovetailed nicely with one another. As Shermer noted, civic and business leaders in postwar Phoenix had benefited handsomely from the opportunities that the war presented. The city's boosters built on

this foundation to attract postwar investments and thus promoted the rise of many large work sites. Those places, however, were not necessarily built to last.[15]

This book fits well on the shelves of business, labor, and western history, but a few of its building blocks come from the social sciences. In the early stages of this project, I found a better sense of direction in the work of American sociologist Charles Fritz, one of the pioneers in the field of disaster studies.[16] As a young army officer in World War II, Fritz evaluated the impact of Allied bombing campaigns over Germany, an experience that forever shaped his academic career. In the early postwar years, the captain-turned-scholar authored several studies that still resonate in the social sciences. In 1961, for example, he offered an oft-cited definition of disasters: "In its most general sociological sense," Fritz wrote, "a disaster is defined as a basic disruption of the social context within which individuals and groups function, or a radical departure from the pattern of normal expectations." To add more meaning to his definition, Fritz described disasters as events confined by notions of time and space that have had "destructive effects on people and physical objects."[17] This definition served my interests well: it was broad enough to encompass a wide variety of workplace tragedies but sufficiently narrow to marginalize those that claimed relatively few lives and had little effect on everyday business operations and work routines.

In the sphere of work, disasters were usually sudden and unexpected. They typically occurred in the blink of eye, a single moment in time. But these events have occasionally played out over the course of days, weeks, or even longer.[18] It took months if not years for devastating droughts to end, for example, which had long-term implications for agricultural workers and their families. Regardless of their duration, however, Fritz and other social scientists have made it clear that disasters have had life spans that passed through three distinct stages: pre-impact, impact, and post-impact. Disaster studies have concentrated most heavily on the latter period, where the "lessons" of these events can be found. In this book, I wanted to focus on all three stages in the life cycles of disasters and thus offer a relatively broad view of the subject matter. This is not to say that disasters have clear beginnings and endings. It is simply to say that they have deep historical roots and long-term implications.[19]

Political scientists have often studied the impact of disasters on public

policy, and this book draws on their work. The studies of John Kingdon proved especially useful. In *Agendas, Alternatives, and Public Policies* (1984), Kingdon described disasters as "focusing events," a phrase that sheltered a set of ideas and assumptions. As Kingdon explained, catastrophic events often bring social problems and politics into sharp alignment, opening "windows of opportunity" to force policy changes. The power of these events, however, depended on how thoroughly the mass media covered them and the extent to which "policy entrepreneurs" influenced post-disaster political domains.[20] Another political theorist, Thomas Birkland, brought these ideas into sharper relief. In *After Disaster* (1997) and *Lessons of Disaster* (2006), Birkland asked why some catastrophes had more focusing power than others. In answering this question, Birkland broke down the basic characteristics of focusing events and showed more clearly how the media and political leaders shaped safety reform movements.[21]

This project also reflects ideas of Paul Sabatier, whose work provided yet more insights into the complexities of policy making. More specifically, Sabatier explained how disasters brought together advocacy coalitions that engaged in debates with one another over proper policy responses. The interactions between advocacy groups, he pointed out, have never been simple or routine. Among other things, they involved careful evaluation of technical information and expert analysis. Like legislators who pushed for policy reforms, advocacy groups demanded details about the potential benefits and costs of new safety proposals. A host of individuals and professional associations provided that information, including prominent scientists and leading research universities. The salient point is that post-disaster policy making involved an assemblage of people and organizations, each of whom had their own outlooks and interests. Through their interactions, they tried to reach mutually beneficial outcomes.[22]

Ingar Palmlund offered another model of policy making that proved inspiring. As an investigator of occupational health and safety problems in her home country of Sweden, Palmlund gained a deep understanding of how advocacy groups searched for political solutions to workplace safety problems. In her work on risk management, she compared the process of setting new safety standards to social dramas. The central plots of these dramas resembled classical tragedies that played out on ancient theater stages. Their story lines centered not only on a crisis but also on the desire to bring order and equilibrium out of turmoil and anxiety. Their characters, Palmlund sug-

gested, fulfilled six generic and sometimes overlapping roles. Risk bearers and risk generators represented the protagonists and antagonists in the dramas and had the most at stake. The former group's advocates pushed for safety reforms while risk researchers gathered information for them. Meanwhile, risk informers kept issues before the audience, and risk arbiters worked to resolve conflicts. In the final acts of these dramas, the plots and characters entered a period of "reintegration," when new safety policies emerged and order was finally restored. This model offered a colorful blueprint for discussing post-disaster policy making.[23]

Social scientists have also helped to contemplate the role of technology in post-disaster policy making. As their studies have shown, safety reform advocates had a better chance of achieving their goals if they could use recent technological advances to their advantage. On many occasions, the availability of promising new safety devices allowed policymakers to gin up support for their policy proposals and to shepherd those proposals through the policy-making process. Conversely, the absence of such devices dampened enthusiasm for policy proposals and thus held up or halted safety reform movements. In some cases, however, the absence of new safety devices prompted policymakers to deploy old technologies in new ways. These ideas lay behind another subtheme in this book: while stressing that new risks accompany innovations, the work also shows that innovations have often been deployed in ways that protect lives and property.[24]

No one will mistake this book for a work of social science. It is not about the strengths or weaknesses of any particular theoretical model, nor does it fully integrate social science concepts into each of its chapters. The book offers no statistics or graphs to explain human and institutional behavior, and it likely overlooks or marginalizes some potentially relevant social science studies. But it does recognize contributions of the social sciences to the field of disaster research, and it suggests how those of us in the humanities can draw on such contributions to make our projects more meaningful and interdisciplinary. Sociologists, political theorists, and other social scientists help us understand *when* and *why* disasters generate policy changes, and *what form* those changes take. They also help to cast the policy-making process as contested terrain, where individuals and groups pursue their own specific goals. Perhaps most importantly, they help with viewing disasters as historical events linked across time and space.[25]

↯

Within a larger frame, this book is about the strange relationship between forces of creation and destruction. Like many historians, I developed an interest in this relationship after reading the works of Joseph Schumpeter, the Austrian-born political economist who famously compared capitalism to a "perennial gale of creative destruction."[26] In *Capitalism, Socialism and Democracy* (1942), Schumpeter described modern economic life as increasingly chaotic and unstable. "Economic progress, in capitalist society, means turmoil," he said so succinctly.[27] More than his contemporaries, Schumpeter saw that capitalism was not a static or stationary mode of production but rather a dynamic economic system that was always changing. He saw entrepreneurs, acting either as organizations or individuals, revolutionizing this system "from within." When Schumpeter spoke of "crumbling walls," "decomposition," and the "evaporation of property," he was referring to the evolutionary nature of capitalism. But he was also urging his peers to stop asking how capitalism administers structures and consider instead how it "creates and destroys them." "As long as this is not recognized," he cautioned, "the investigator does a meaningless job."[28]

Business and economic historians continue to wrestle with the implications of these words. Mark Wilson's *Destructive Creation* (2016), an investigation of America's World War II war mobilization campaigns, offers the most relevant evidence. In brief, Wilson's work showed how coalitions of public agencies and private businesses forged one of history's most powerful, inexorable forces—a massive arsenal of highly destructive killing machines. "A giant capitalist economy was harnessed for the purpose of annihilating its enemies," Wilson wrote. This "creation," the historian concluded, "owed as much to socialism as it did to capitalism." Only the federal government had the resources and organizational skills to create such a giant complex of industrial facilities, supply chains, and agencies. Ironically, destruction abroad hastened the construction of ever more deadly weapons at home.[29]

Kevin Rozario's *The Culture of Calamities* (2007) offers another good example of a recent study that investigates links between creative and destructive forces. Exploring catastrophes in American history since colonial times, Rozario has shown that Americans have long been bombarded with "spectacles of destruction." The historian asked, Why are these spectacles so ubiquitous? And why are we so enamored of them? Rozario found his answers in our growing anxieties about the future and in the way the mass media has turned calamities into entertainment. But he also attributed our fascination

with disasters to the belief that these events "make things happen."[30] In one case after another, Rozario argued, catastrophic events have stripped away the artificialities of daily life and motivated people to take action, thereby creating opportunities to make the world safer. As a result, we have often come to see disasters, unconsciously perhaps, as "instruments of progress."[31]

The idea that calamities have engendered progress has a long history. In Western cultures, stories of mythical gods, great floods, and spiritual resurrections portray cataclysms as mere preludes to periods of recovery and restoration. Similar ideas permeated religious thinking in colonial America. Puritan leaders in Massachusetts Bay Colony warned repeatedly that God might one day destroy the earth, but not without creating a better world for those who heeded his ways. During the nineteenth century, European intellectuals like Charles Darwin, Sigmund Freud, and Karl Marx suggested that all things "modern" had their origins in great conflicts and struggles for survival.[32] The German philosopher Friedrich Nietzsche went so far as to say that chaos itself was a creative force and that "good" and "evil" were possibly one and the same, like twins joined at the hip. In subtle and incalculable ways, these and other deep thinkers embedded a haunting image in the public consciousness—that of two opposing forces moving through modern life, one of which was destroying everything in its path while the other created the world anew.[33]

By the twentieth century, this double-fisted image figured prominently in works of fiction, among them classics of Western literature. Consider, for example, the main idea in Robert Louis Stevenson's *Strange Case of Dr. Jekyll and Mr. Hyde* (1886). Stevenson's story of a doctor who brought to life his "second self" implied that we as individuals are essentially two people, one of whom is insidiously evil and destructive, the other sensitive, caring, and even creative.[34] Other novelists of the period relied on concepts of chaos and destruction to create utopian visions of the future. Ignatius Donnelly's *Caesar's Column* (1890), for example, depicted a violent world from which its protagonist escaped and built a peaceful agrarian society.[35] Donnelly, a founding member of the short-lived Populist Party, had delivered a similar message in *Ragnarok: The Age of Fire and Gravel* (1883), a story about a comet's impact on an advanced ancient civilization. "So far as we can judge," Donnelly concluded in that book, "after every cataclysm the world has risen to higher levels of creative development."[36]

This idea slipped into further recesses of the mind with the rise of the

motion picture industry. As film historians have pointed out, early filmmakers often staged train wrecks, house fires, and other acts of destruction to titillate audiences. But they sometimes juxtaposed these acts with scenes of creation. The industry's first slapstick comedy, *An Interesting Story* (1904), offers one of many good examples. The short British film showed a happy-go-lucky chap wander into the path of a moving steamroller that flattened him like a pancake. When two passing bicyclists came across the man's dead body, they grabbed their bicycle pumps and filled him up with air, restoring the fellow to life. Thomas Edison's film studios made a more memorable contribution to this genre in 1910, when it produced the first screen adaptation of Mary Shelley's famous *Frankenstein*. As readers no doubt remember, Dr. Frankenstein eagerly harnessed science and technology to create the "perfect human being," but he inadvertently made a "miserable monster." This "wretch," or "creature," threatened to destroy everything the good doctor held dear. Creation and destruction, ruin and renewal—theater audiences on both sides of the Atlantic enjoyed this theme, and filmmakers increasingly integrated it into their work.[37]

As I sifted through this material, I started to see my subject matter in a new light. Patterns of havoc and reform began to look like two strands of a single cord, or links in a long Byzantine chain. Disasters no longer seemed like isolated events or accidents that happened under rare and unusual circumstances. They appeared to be part of a larger phenomenon, as did any reforms they explained. Workplaces themselves looked more like ephemeral objects than solid structures, and the people within them were merely passersby. All things were caught up in the orbit of two forces that kept turning the world upside down, not just here and there, but always and everywhere. This outlook, a perspective on the past, is deeply rooted in literary and artistic endeavors. It assumes that nothing is stable or safe, but it also presumes that people and communities are resilient, built to rebuild.

This perspective came into sharper focus when I encountered the work of the late Marshall Berman. In *All That Is Solid Melts into Air* (1982), the philosophic Berman waxed eloquent about "the experience of modernity." Like Schumpeter, Berman described the world as a place of invariable turmoil.[38] While reflecting on the changing landscape of his hometown, New York City, the self-described modernist found himself surrounded by bulldozers, giant cranes, and "hundreds of workers in their variously colored hard hats." Buildings were constantly crashing down and springing up, and

in the process transforming patterns of life and culture. Berman envisioned himself in a "tragedy of development" and "a maelstrom of perpetual disintegration and renewal." But Berman was no pessimist or fatalist. Rather, he was a humanist. He believed that modern societies have an innate ability to restore themselves, what he called "an infinite capacity for redevelopment and self-transformation." As he concluded, "I believe that we and those who come after us will go on fighting to make ourselves at home in this world, even as the homes we have made, the modern street, the modern spirit, go melting into air."[39]

Modernity is an elusive concept, and I have not tried to corral it in this book. At the heart of this work, however, are a few basic assumptions about the character of our modern era. The work suggests that our ever-changing world is moving in ways we can never fully understand. It highlights the historic roles of technical advances, new commodities, and large occupational groups while also accentuating the significance of public institutions and policy making. Perhaps more importantly, it suggests that we "moderns" live and work in a world filled with dangers—the kinds that have enormous destructive potential. Not surprisingly, the greatest of these dangers are our own making. Most were inconceivable before the Industrial Revolution and even unknown after it had largely run its course. Any study of recent disasters necessarily speaks to the growing complexities of modern life and to the ways in which societies have tried to manage them.[40]

1

Disasters at Work

A Brief Background

Along with the growing capacity of technical options grows the
incalculability of their consequences.

Ulrich Beck, *Risk Society*

Major industrial and technological advances have often performed
like a magnet, pulling people to new work environments where they produced a constellation of new goods and services. But those same advances
have also created a host of hazards, occasionally generating large-scale catastrophes that destroyed places of employment and upended the lives of
everyone within them. This paradox was especially clear in the United States
from the mid-nineteenth through the early twentieth centuries, when the
nation became the world's leading industrial power. While steam-driven engines, threshing machines, and other extraordinary innovations transformed
the American economy, factories burned, trains derailed, and mines collapsed. Unfortunately, neither employers nor public leaders seemed overly
concerned about this painful, somber side of material growth. Common-law
doctrines protected the employers against lawsuits, and the federal government generally stayed out of the internal affairs of private businesses. Gradually, however, the growing scope and lethality of these events prompted
American policymakers at all levels of government to find new legal and
organizational solutions to safety problems. By the time the United States
entered World War II in 1941, the nation had built up an impressive regulative framework of codes, laws, and agencies designed to make workplaces
safer. These various reforms benefited the general public as well as employers and their employees.

This chapter explores these patterns and trends. Though not a comprehensive survey of the subject matter, it identifies more than a dozen prewar disasters that wrecked large work sites and hastened the rise of salutary safety regulations. Along the way, it sheds light on a few less newsworthy tragedies that also promoted safety reforms. The purpose of the chapter is not to summon horror stories from the distant past, some of which are already fairly well known. It is instead to bring the two-sided character of industrial and technological change into sharp relief, and thus to provide an appropriate backdrop for the book's five case studies of postwar disasters. The chapter's montage of sketches and scenes reveals patterns of ruin and renewal in a wide range of industrial settings. In the century before World War II, an era of remarkable innovation and expansion in all sectors of economic life, these patterns were hard *not* to see.

The chapter is organized along industrial and occupational lines. It begins in the realm of transportation, at a time when steamboats and railroads shuttled growing numbers of people and commodities across the country. It then moves into the fields of mining and manufacturing, which by the turn of the century were America's leading engines of job growth. The journey ends in the new century's expanding service sector, in places of employment such as theaters and schools. At that point, the ubiquity of workplace disasters in American history should be evident. It should also be clear that by the outbreak of World War II, these tragic events had created a cornucopia of safety reforms. A strange image of creative and destructive forces barreling through modern life might also have come into view.

The problem of workplace safety is neither novel nor uniquely American. Salt mines, sailing vessels, and other ancient work environments had proven unsafe. But the rise of highly capitalized firms filled with dangerous machines and dozens of wage earners dramatically increased the likelihood of workplace injuries. The problem first became obvious in England, the cradle of the Industrial Revolution. By the mid-nineteenth century, English factory inspectors had reported thousands of workplace accidents, hundreds of which resulted in deaths or amputations. Local newspapers presented graphic accounts of these accidents, detailing how workers suffocated in coal mines or lost fingers, hands, and arms in textile factories. These stories generated widespread sympathy and indignation, especially when the victims were women or children. Unfortunately, disaster victims and their families had to

rely on the charity of their local communities for assistance because factory owners and mine operators typically provided little if any compensation for work-related deaths and injuries.[1]

In the United States, where industrialization took hold more slowly, the deadly nature of new places of employment first became apparent on the nation's rivers, where steam-powered paddle wheel boats exploded with astonishing regularity. Historians have carefully tracked these explosions, sorting them by decade, death toll, and even the rivers on which steamboats blew up. Our enduring interest in these events is well justified. By the 1840s, there had been more than one hundred of these explosions in the country, claiming some six hundred lives. Rarely have such tragedies been defined as a labor problem, though many of their victims were employees. Steamboat crews were far larger than commonly assumed. In the mid-nineteenth century, the average American steamboat had a crew of two dozen workers, and larger showboats catering strictly to passengers employed crews of eighty or more. These workers fell into two general categories: deckhands and cabin crews. The former included pilots, engineers, and boiler stokers; the latter consisted of cooks, waiters, barkeeps, cabin attendants, chambermaids, and even musicians. The people who filled these positions were an unusually diverse group. In the lower Mississippi Valley, many of the deckhands and cabin workers were slaves.[2]

The great satirist Mark Twain offered a fascinating account of a steamboat explosion in *The Gilded Age* (1873). Though basically a story about greed and corruption, the novel went into great detail about a "race" between two steamboats on the Mississippi River. The contest turned tragic when an overworked steam engine on one of the boats erupted like a volcano. "And then there was a booming roar, a thunderous crash," Twain wrote. "The whole forward half of the boat was a shapeless ruin, with the great chimneys lying crossed on top of it." The tomfoolery of the boat's captain killed more than a hundred people, yet no one was charged with a crime. "A jury of inquest was impaneled, and after due deliberation and inquiry they returned the inevitable American verdict which has been so familiar to our ears all the days of our lives—'Nobody to blame.'"[3] That, too, was a sign of the times. In the United States no less than England, judges and courts had yet to hold employers accountable for the health and safety of their employees and customers. State legislators worried that costly fines and safety regulations would only hamper industrial growth, and thus they tolerated a great degree

of risk in the business world. The question of fault in these explosions was also difficult to assign with any real sense of certainty. Evidence of negligence, if it could be found, proved unpersuasive.[4]

What may have inspired Twain's story was the *Moselle* tragedy of 1838. Named after a tributary of the Rhine River, the *Moselle* was one of the newest, finest, and fastest steamboats on the Ohio River, where it ferried goods and passengers between river ports. The boat's owner and captain, Isaac Perrin, enjoyed a reputation for swiftness; he kept his boat's engines under constant pressure, even as passengers boarded. On April 25, Perrin was apparently racing a competitor on a journey from Cincinnati to St. Louis when his engines unexpectedly detonated like barrels of gunpowder, shredding the magnificent vessel and killing about 150 people, including Perrin and nearly a dozen of his boat's 30 crew members. Survivors, many of whom were badly injured, struggled to the river's shores, where they began calling and searching for loved ones. Widely reported in newspapers of the time, the tragedy convinced reluctant policymakers to rethink their minimalist approach to regulating business activities. Only a month after the tragedy, the US Congress passed the 1838 Steamboat Act. The act required steamship companies to comply with a slew of new safety regulations, among them the requirement for steamboat operators to have their ship's boilers inspected every six months. More importantly, it suggested that boiler explosions should be viewed as evidence of negligence in future lawsuits involving steamboat accidents.[5]

The federal government's response to the *Moselle* tragedy highlights an underlying premise of this book: that the forces of creation and destruction have often traveled together, seemingly hand in hand. In 1838, these forces came into sharper focus. Until then, the federal government had not fully acknowledged the dangers of steamboat travel, nor had it taken such significant steps to regulate an obviously dangerous industry. Proposals to make steamboats safer had faced strong resistance from a Congress opposed to interfering in the internal affairs of private businesses. Even President Andrew Jackson had failed to convince lawmakers to address the steamboat safety problem.[6] But the 1838 disaster changed this state of affairs. It stirred up what political theorist John Kingdon has called the "primeval soup" of policy making. To most Americans, including federal legislators, traveling on steamboats was something tangible, personal, and easily imagined. The idea that these vessels could suddenly erupt like volcanic craters genuinely

frightened the public, to whom lawmakers had to answer. Congress consequently created something new—a law designed to bring a sense of stability to an ever more unstable situation.[7]

Unfortunately, the Steamboat Act of 1838 proved woefully inadequate. Among other things, the act failed to ensure that safety inspectors were well trained or even honest; in fact, some were downright incompetent and easily bribed. Riverboat disasters of the 1840s shone a glaring light on the problem. More than fifty riverboats exploded during that decade, claiming more than five hundred lives. Bold headlines and sensational stories in newspapers brought these events to public attention. A journalist's account of an explosion on the lower Mississippi in 1843 exemplified the news coverage. The account first appeared in the Louisiana *Chronicle*, describing a horrifying scene: "huge beams of timber, furniture, and human beings in every degree of mutilation were alike shot up perpendicularly many hundred fathoms in the air." "On reaching the greatest height, the various bodies diverged like the jets of a fountain, in all directions—falling to the earth, and upon the roofs of houses, in some instances as much as two hundred and fifty yards from the scene of destruction."[8] Other accounts of the tragedy used words like "scalded," "mangled," and "crushed" to describe the dead. Some even reproduced graphic illustrations that showed people bleeding, screaming, and drowning. Many also featured heartfelt editorials that called on Congress to find new ways of reducing the risks of steamboat travel.

A series of new riverboat disasters in the late 1840s and early 1850s put more pressure on Congress to act. Newspapers devoted front-page coverage to these calamities and often printed follow-up stories about them. Two of these tragedies—the destruction of the *Anglo-Norman* and the *Saluda*—proved particularly newsworthy. The beautiful new *Anglo-Norman* exploded in 1850, just after departing from New Orleans on a pleasure trip up the Mississippi. Nearly one hundred people lost their lives in the tragedy, several of whom were relatively wealthy, well-known citizens. When the *Saluda* blew up two years later, in a narrow channel on the Missouri River, another hundred people perished. In this case, dozens of victims were recent Mormon immigrants on their way from England to Salt Lake City, Utah. By the time the *Saluda* exploded, Congress had already begun debating new proposals to strengthen federal steamboat regulations. The debate ended soon after the accident, when Congress passed the 1852 Steamboat Safety Act.[9]

Like modern forms of government regulation, the Steamboat Safety Act

EXPLOSION AND BURNING OF THE OHIO RIVER STEAMER "MAGNOLIA," NEAR CINCINNATI, OHIO, March 18, 1868.

Engraving of the *Magnolia* ablaze on the Ohio River near Cincinnati, March 17, 1868. Steamboat explosions were among America's earliest industrial disasters, and they plagued the nation throughout the nineteenth century. About forty people perished in this tragedy, including the boat's captain and several members of his crew. Courtesy of the Kentucky Historical Society

of 1852 established a semiautonomous agency—the Board of Supervising Inspectors—to oversee business activities in a specific area of the national economy. To put it briefly, the board took over the steamboat inspection process. Its nine members, who earned annual salaries for their work, set new licensing standards for steamboat inspectors and reserved the right to inspect steamships without notice. They could also call hearings to determine whether steamship operators had been negligent in their conduct and to issue public notices identifying commercial vessels that they deemed unsafe. Though the act had its own limitations, it nonetheless achieved its main

purpose. Five years after its implementation, the annual number of boiler explosions on American rivers had dropped by more than 75 percent, which no doubt saved more than a thousand lives.[10]

By midcentury, railroads were rapidly replacing steamboats as the nation's primary carrier of people and goods. Like steamboat disasters, train wrecks captured national attention and thus encouraged employers and public leaders to address workplace safety problems. In *Death Rode the Rails* (2006), historian Mark Aldrich spoke pointedly to the close relationship between disasters and safety regulations. As Aldrich noted, rail safety had become a matter of great concern in the mid-nineteenth century after terrible train wrecks in both Europe and the United States.[11] An 1842 derailment in France near the city of Versailles had left a particularly deep impression on the public imagination. The wreck occurred on the birthday of King Louis Philippe, and many of its victims had participated in a national celebration of the event at the Palace of Versailles. A broken axle forced the train's front locomotive off the track, and the second locomotive and three carriages filled with passengers then crashed along the roadway, killing more than fifty people and injuring many others.[12]

In the United States, the rail safety problem grew clearer on May 6, 1853, when a train traveling from New York City to Boston plunged off an open drawbridge and into Connecticut's Norwalk Harbor. The derailment killed about fifty people and injured many more, including several members of the train's crew. Newspapers suggested the train's engineer missed a crucial warning sign as he approached the bridge, but regardless of whether he alone caused the crash, the accident terrified the general public. How could a careless mistake have claimed so many lives? What could be done to prevent a similar tragedy? In response to such questions, Connecticut passed legislation requiring all trains in the state to stop before crossing drawbridges. The law, however, was a relatively ineffective way of addressing a serious safety problem. It not only delayed rail travel but also imposed an annoying burden on train engineers. More importantly, it revealed a lack of knowledge about the potential of recent advances in the science of traffic management and railroad signaling. Yet the legislation seemed reasonable at the time. It offered policymakers a quick and inexpensive way to address public concerns.[13]

Press coverage of these events skewed heavily toward the impact of the disasters on rail passengers and thus obscured the extent of victimhood

among trainmen. As historians of the railroad industry have made clear, the number of trainmen killed in train wrecks was no small matter. Though they made up only one-fifth of the railroad industry's workforce, trainmen accounted for nearly half of all its deaths and injuries. By the 1880s, more than one thousand of these men died on the job annually, and many more suffered serious injuries. These figures seem even more startling considering that train crews were relatively small. Freight trains typically employed only six men. An engineer, fireman, and a head brakeman worked at the front of freight trains, and a conductor, brakeman, and a flagman worked from the rear. Passenger trains could carry nearly a hundred travelers at the time and thus required the employment of other groups of workers, including sleeping-car porters, baggage handlers, and other trainmen who tended to the needs of passengers. Like steamship explosions, then, rail disasters upended the lives of countless numbers of people at work.[14]

The lethality of rail disasters grew more alarming as the nation's rail system expanded. One of the most disturbing of these events occurred on the evening of August 26, 1871, when an express train heading north from Boston, Massachusetts, slammed into the rear of a smaller train at a rail station about five miles from the city, in the popular beach community of Revere. The collision killed twenty-nine people and injured nearly sixty others, and it was both horrendous and inexcusable. The larger train's locomotive burrowed its way into the smaller train's rear passenger car, where sixty to seventy people were sitting. Simultaneously, red-hot coals and scalding steam from the locomotive's busted smokestack and boilers scattered through the area and mixed with oil spilling from the passenger car's burning lamps. A fire subsequently engulfed the entire crash site. The horror of the moment was hard to imagine, though news reporters did all they could to re-create it.[15]

The tragedy in Revere provoked outrage. As the public soon learned, poor communication between men in charge of the trains and the station largely accounted for the crash. The public's fury was swift and soon spread statewide. One well-known social reformer, Wendell Phillips, accused railroad companies of "deliberate murder."[16] The newly established Massachusetts Railroad Commission, one of America's first state regulatory agencies, conducted a lengthy investigation of the wreck and concluded that no laws had been violated. But the commission's chief policymaker, Charles Francis Adams, used the tragedy to mount an unusually successful reform move-

ment. As the grandson of President John Quincy Adams and a former military officer in the Civil War, Adams was widely respected and well connected, and his condemnation of the Revere accident reverberated far and wide. In its aftermath, Adams played the role of a *policy entrepreneur*. He spearheaded a public relations campaign that shamed eastern railroads into standardizing their operating procedures and adopting newly patented rail safety appliances, including advanced braking systems, better traffic signals, and railcars that were more capable of withstanding collisions. In the words of business historian Thomas K. McCraw, these developments amounted to a "safety revolution."[17]

Another tragic event of the 1870s exposed the link between rail disasters and safety reforms. On December 29, 1876, in frigid conditions just after dark, a large passenger train carrying 159 people plunged off an iron bridge spanning the Ashtabula River in northeastern Ohio. One of the train's two locomotives and several passenger cars dropped like stones, piling up in the river and along an adjacent ravine. In the midst of this mayhem, the train's coal-burning stoves scattered through the falling wooden passenger cars and turned the crash site into a massive bonfire. The derailment and the fire it sparked claimed some eighty lives and left sixty people badly injured. Among the dead was one of the nation's most acclaimed singers and composers, Philip Bliss, whose name appeared in most news accounts of the tragedy. Largely overlooked in these accounts were the trainmen who lost their lives that night. When their workplace went off the rails, porters, baggage handlers, and other members of the crew departed this world.[18]

Accident investigators blamed the Ashtabula derailment on a broken car axle, but they also acknowledged that the bridge on which the derailment occurred was fundamentally flawed. Its creator, Amasa Stone, was a former railroad president with no formal scientific training. The bridge he designed was essentially a relic of a bygone period when trains were lighter, shorter, and slower. The realization that bridges could be so fundamentally flawed motivated a string of safety reforms. Railroad companies, which faced new legal liabilities as a result of such tragedies, began working with state railroad commissions to review the structural integrity of railroad bridges everywhere, a huge and expensive undertaking. The derailment consequently encouraged engineers to develop new ways of building and inspecting bridges, and of measuring the effects of weather and traffic on the performance of

bridges. Furthermore, the tragedy encouraged manufacturers to develop safer heating systems for passenger cars and to replace wooden cars with fire-resistant models made of steel. Though not formal laws, these safety measures nonetheless proved beneficial.[19]

Train wrecks continued to inspire safety reforms well into the twentieth century, but by 1900 the likelihood of dying in rail accidents had already begun to decline. This trend partly reflected the effectiveness of safety reforms. In response to public anger over rail-related deaths and injuries, state officials and railroad companies had found new solutions to rail safety problems. For example, the companies had invested in sturdier rails, better roadbeds, and safer locomotives. Meanwhile, more states had set up public commissions to investigate accidents and promote new rail safety technology. The US Congress finally joined this movement by passing the 1893 Railroad Safety Appliance Act. Signed into law by President Benjamin Harrison, who had long called for such legislation, the act essentially federalized state laws and trade group policies that promoted the adoption of new inventions like automatic car-coupling devices and better braking systems. The federal government's response to the safety problem was less than forceful, however. It gave railroads five years to meet the newly imposed safety standards and then extended the deadline for another two years. The response was nonetheless effective. By the turn of the century, new safety regulations had no doubt saved the lives of thousands of people, including hundreds of railroad workers.[20]

On the eve of World War I, however, workplace fatalities in the United States had reached an all-time high. According to the US Bureau of Labor Statistics, about twenty-three thousand Americans died on the job in 1913 (out of a workforce of roughly thirty-eight million).[21] Such figures impelled the nation's labor commissioner, Charles Neill, to slam industrialists before the National Safety Council, a newly formed organization of business and civic leaders interested in promoting public safety. Speaking at the organization's second annual meeting, Neill suggested that no other nation in the world had inflicted such pain on its workforce: "I doubt if there is a commercial nation today, laying any claim to an elementary civilization, that has been maiming and mangling and killing those who attempt to earn their bread in the sweat of their faces with as little apparent regret and as little thought as we do," he told the council. From Neill's perspective, the nation had

an ethical duty to protect its workforce. The new National Safety Council, Neill hoped, could stop what he called the "slaughter of the unoffending."[22]

The word "slaughter" recalled recent coal mine disasters in the Appalachian Mountains and other parts of the East and Midwest. At the turn of the century, when there were no federal mine safety laws, coal mining was as dangerous as working on railroads. Hazards were everywhere. The ground above coal mines was usually soft, flat-lying rock prone to collapse, and mines themselves were often not properly reinforced. The use of dynamite in mining posed the most obvious danger, but open-flame lamps, rock-cutting machines, and other accoutrements of the trade sparked catastrophes as well. Simply walking through mines often proved fatal. These work environments were not only dark but also narrow, and their ceilings were only five or six feet high. Carloads of coal often appeared out of nowhere, pulled by mules or small locomotives. Wherever machines did the hauling, high-voltage wires stretched through passageways and sometimes touched up against miners' damp hats and clothing. To cap it all off, many miners were young and inexperienced, and mistakes could kill. Employers typically paid miners by the ton and thus encouraged them to work faster, without interruption or rest. Statistics from the Labor Bureau underscored the problem. By 1910, mining accidents were killing more than two thousand miners annually. In 1907 alone, three thousand miners died in work-related accidents.[23]

The 1907 tragedies included the nation's worst mining accident. On December 6, near the small community of Monongah, West Virginia, mines owned by the Fairmont Coal Company ignited and collapsed. With one thunderous boom, 360 miners died, many of them recent immigrants from southern and eastern Europe. Though investigators never determined the exact cause of the explosion, they surmised that a miner's lamp ignited methane gas or coal dust in the air and that the resulting blast brought down wooden timbers supporting the mines while also destroying ventilation systems, leaving the miners buried alive.[24] Whatever the case, Monongah was more than just the "worst" occurrence of a needless tragedy. It showed, more plainly than ever, just how deadly workplace accidents had become. The West Virginia blast wiped out workers by the hundreds, not slowly over time but suddenly and without warning. It also flung their families into states of deep despair. About 250 women lost their husbands in the 1907 tragedy, and potentially 1,000 children lost their fathers. This was a disaster in the true

Weary miners at Monongah, West Virginia, December 8, 1907. Beyond this entrance, 360 miners lost their lives in the nation's worst mining disaster. Courtesy of the US Bureau of Mines Collection, Mine Safety and Health Administration Technical Information Center and Library, National Mine Health and Safety Academy

sense of the word. Resembling a major earthquake or hurricane, it disrupted an entire community and carried great social and economic costs.[25]

Coming on the heels of other mining disasters, the tragedy at Monongah convinced the federal government to take its first steps toward regulating the mining industry. Only six months after the tragedy, Congress authorized the US Geological Survey to map the nation's natural resources and to set up stations for the purpose of investigating mine explosions. At the same time, the State Department appealed to Germany, Great Britain, and Belgium (where fatality rates in mining were relatively low) for help in minimizing such tragedies. In response, mine safety experts from those nations led a two-month investigation of American mines and made several recom-

mendations. Most importantly, they called on public officials to regulate the use of explosives, lamps, and electrical equipment in mines, and to see that coal companies kept all roadways within their mines free of loose coal because it easily ground into combustible dust.[26]

As federal officials reviewed these recommendations, more mining disasters rocked the nation. The most appalling occurred near the small town of Cherry, Illinois. On November 13, 1909, a fire swept through mines owned by the St. Paul Coal Company, killing 260 miners. It started after miners apparently parked a cart filled with hay for mules close to a kerosene lamp. Small fires were common in mines, and miners could typically extinguish them quickly. In this case, however, the combination of kerosene from the lamp along with the rush of fresh air coming from a nearby airshaft fanned the fire's flames and soon consumed the overhead timbers supporting the mine. The fire then intensified and spread out of control. Supervisors sealed mine entrances to starve the fire of oxygen. By then, however, the blaze had consumed everyone who was still beneath the ground. The men who perished in the disaster left behind 160 widows and nearly 450 children, most of whom were recent immigrants from Europe.[27]

Some of the most horrifying of these tragedies happened in the American West, during a time when the nation's thirst for coal and copper brought new capital resources and workforces to western territories. Western mines had caved in, caught fire, and exploded since the great gold and silver rushes of the nineteenth century, but only a few of these accidents claimed more than a dozen lives. By the 1920s, such deadly events were all too common in the West. An accident in 1913 near Dawson, New Mexico, killed 260 miners. In 1917, a disaster near Ludlow, Colorado, killed 120 miners, and another in the copper-rich mines around Butte, Montana, claimed 170 lives.[28]

Labor leaders generally blamed these explosions on the greed of mining companies that seemed bent on extracting the highest possible profits from the production process. None of those leaders was more critical of mine operations than Bill Haywood, head of the militant Industrial Workers of the World (IWW). Haywood himself worked in western mines for more than a decade and had an all too intimate understanding of the dangerous nature of mining activities. In addition to suffering a debilitating hand injury in a mine, he also had the sickening experience of finding the lifeless, disfigured bodies of his fellow workers inside mines. "Big Bill" recalled these experiences in his 1929 autobiography: "a slab of rock fell and crushed the life out

of him," he said of one miner. Another, he wrote, "had his entire face blown off." It was understandable that Haywood's guiding principle was "an injury to one is an injury to all." It was also more than mere rhetoric that the IWW promised to give workers "control of the machinery."[29]

These various tragedies and the anxieties they generated resulted in a rash of investigations, lawsuits, and coroner inquests, which in turn prompted policymakers to take mine safety more seriously. It was no coincidence that Illinois strengthened safety regulations after the Cherry Mine disaster, or that other states soon followed suit. More telling was Congress's decision in 1910 to create the US Bureau of Mines. Though Congress limited the bureau's power to matters of research and investigation, the agency nonetheless worked diligently to make mining safer. Like the nation's earliest railroad commissions, the bureau distributed crucial information about the possibilities of new safety technologies and discussed safety problems with industry leaders and public officials. Yet the creation of the bureau was a relatively weak response to the mine safety problem, and mining disasters continued to plague the nation for years. When the United States entered World War I in 1917, the Bureau of Mines still had no authority to inspect mines or force employers to change their practices.[30] The global conflagration turned America's mineral resources into vital war commodities and thus ratcheted up the pace of production. That only worsened conditions in mines. Not until 1941, after yet more mining disasters, did Congress at last give federal inspectors the power to enter mines and report violations of local safety standards. Even then, the well-being of miners remained largely in the hands of local and state authorities.[31]

To be clear, catastrophic events have not always translated to safer places of employment. In mining, it took a wave of terrible tragedies to trigger even the slightest political reactions. The slow pace and limitations of mining regulations mirrored circumstances that distinguished this area of industrial activity. For starters, mining took place in relatively remote parts of the country and below the ground, making it relatively difficult for the media to cover and sensationalize mining disasters. Drawings and descriptions of these disasters were simply less captivating than those of exploding steamboats and mangled railcars.[32] Perhaps more importantly, mining disasters killed only miners, not middle-class citizens out for pleasure cruises or popular singers and composers on tour. The victims were mostly recent immigrants with little money, many of whom were viewed with indifference if not

prejudice by the general public. Mining disasters consequently never generated as much shock and outrage as their counterparts in transportation, which allowed policymakers to move more slowly in governing risks to workers' health and safety. The power of mine owners in the policy-making process also retarded the pace of safety regulations. Even the militancy and solidarity of miners were not enough to make the owners support sweeping, compulsory legislation designed to address safety problems. Like employers in other areas of industry, mine owners saw proposed safety regulations as threats to their prerogatives as employers, and they vigorously resisted government efforts to intervene in their business affairs. In mining communities the owners had considerable clout.[33]

The history of steelworkers has its own place in this story. Though steel mills never exploded or collapsed like steamboats and mines, fatality rates in these massive, noisy places were nonetheless staggering. The rates at Illinois Steel, a company that stretched across two miles of land on the shores of Lake Michigan, illustrate the general patterns. In 1906 the company employed about six thousand men, of which nearly fifty died on the job. More than five hundred of the company's other employees suffered physical injuries at work, and more than a few of those injuries left workers with permanent disabilities. These figures were similar to those in plants around Pittsburgh, Pennsylvania, home to six large steel companies. About two hundred steelworkers in the Pittsburgh area died annually during this period, and more than a thousand others suffered work-related injuries. Giant furnaces in steel mills exploded with regularity, which accounted for a quarter of these deaths and injuries. But furnace explosions were only part of a larger problem. These jobs were unusually dangerous. Steelworkers toiled over boiling vats of liquids and under giant cranes carrying hot metals. They dodged small "dinky trains" loaded with hot steel and heavy "skull crackers" that dropped from the air to break up iron ore. They filled big ladles with molten metal and grabbed hot steel ingots with large pincers.[34]

By 1910, a variety of reform-minded Americans had drawn widespread attention to these dangers. Among the most important was Crystal Eastman, whose pioneering study of industrial life in the Pittsburgh area proved inspiring. Eastman's description of job-related accidents showed exactly how steelworkers lost their lives in mills. Her story of the death of Angelo Guira, a seventeen-year-old in the mills, was one of many that rang poignant. As

Eastman explained, Guira had worked near a giant furnace where "welders" with pincers placed red-hot bars of steel on moving rails that snaked through the mills. Guira's job was simply to pull levers on machinery that transformed the bars into tubes. While carrying out this task one day, a tube unexpectedly bounced out of its trough and straight into the young man. The sizzling object was as lethal as a medieval war hammer, and it felled Guira instantly.[35] In another case, a steelworker was standing over an open tank of molten metal "pickling" coils of wires when he lost his footing and fell into the scalding broth. Still another lost his life when an overhead buggy of steel fell and crushed him. Just before the accident, a foreman had yelled at the man to vacate the area, but unfortunately the man did not understand English.[36]

Sadly, steel mills did little more for their injured employees than pay for funeral expenses. They justified their inaction on common law principles. In the nineteenth century, judges and courts had set legal precedents that helped employers avoid liability for workplace accidents. Their rulings generally hinged on the "fellow servant doctrine" that suggested workers willingly assumed the risks of working and were basically responsible for the negligence of their coworkers. Within this legal environment, if "masters" could prove that "servants" had caused their own injuries, then they had no obligation to compensate the injured parties. Employers frequently eluded lawsuits altogether because their employees had neither the time nor the money to sue them. Many industrial workers in the Pittsburgh area were illiterate or did not speak English. Nonetheless, in highly capitalized places like steel mills, the mere threat of lawsuits represented a problem for management. When juries addressed the question of innocence in work accidents, especially those involving large corporations, they often sided with employees, and they occasionally made extremely generous awards.[37]

The carnage among Pittsburgh's steelworkers reflected terribly on their employers, including Andrew Carnegie, the nation's richest steel magnate. Carnegie's own mills near Pittsburgh were as dangerous as any in the area and thus raised the question of how such a wealthy and respected businessman could tolerate these conditions. Was America's great steel master heartless? Had he no shame? Apparently, such questions troubled the "self-made" man. Before Carnegie sold his mills to the US Steel Corporation in 1901, he created an endowed fund for the benefit of his injured employees. Widely applauded, the Carnegie Relief Fund was one of the first of its kind—it of-

fered predictable compensation packages for employees killed or injured in the line of work. More specifically, it offered most widows of steelworkers who died on the job a year of their husband's salary, or roughly four hundred to five hundred dollars (about twelve thousand to fifteen thousand dollars in 2020); it also provided an extra hundred dollars (three thousand dollars in 2020) for each surviving child below the age of sixteen. Furthermore, the fund provided compensation to families of single men who died at work (but only if the families could prove they received regular support from the deceased).[38]

The Carnegie Fund must be understood as a specific response to a changing business environment. By the turn of the century, states had already begun restricting employers' use of common law defenses for job-related accidents, arguing that employers had a moral responsibility to make work sites reasonably safe. Faced with new legal liabilities, employers in all areas of the economy found new ways to lower their accident rates. US Steel was one of the leading actors in this story. After assuming control of Carnegie's mills near Pittsburgh and Illinois Steel's mills around Chicago, the giant corporation took several significant steps aimed at reducing workplace fatalities and injuries. It placed guardrails and warning signs around dangerous machinery, for example, and replaced old furnaces with newer models that were less likely to explode. It also employed teams of safety inspectors who visited plants and made suggestions for making them safer, as well as formed committees to collect and analyze data detailing the circumstances surrounding workplace accidents.

The central figure behind this safety movement was Elbert Henry Gary, who headed US Steel from 1901 to 1927. Though often viewed as a heartless capitalist, Gary had a genuine interest in workplace safety issues. In the 1880s, he worked as a personal injury lawyer representing corporations as well as individuals in cases involving job-related accidents. In the 1890s, he served as general counsel for Illinois Steel, where he gained a deep understanding of the company's safety problems. At US Steel, Gary put together a coalition of managers and safety inspectors for the purpose of improving the safety of mills and in turn invested heavily in safety engineering and education. According to historian Awren Mohun, who studied the industrial safety movement of these years, Gary's investments paid off. Accidents declined markedly at US Steel under his leadership, and payments to injured employees declined accordingly. Gary's accountants estimated that US Steel

saved nearly two dollars on every dollar it spent on workplace safety. The estimate sent a powerful message to other large corporations: "safety first" campaigns were not only humane responses to an increasingly dangerous world of work, but also a cost-effective strategy for dealing with workplace tragedies. Once again, out of tragedies came reforms.[39]

Disasters in other areas of manufacturing have a longer history than often recognized. Several of these events rattled the nation in the late nineteenth and early twentieth centuries. One of the first and most horrific occurred in 1860, in the rapidly industrializing town of Lawrence, Massachusetts. On a cold day in January, the five-story Pemberton Textile Mill collapsed without warning, instantly killing many of the mill's six hundred workers and leaving others trapped in the ruins. Oil lanterns tumbled across cotton supplies and wooden floors, sparking fires that foiled rescue efforts. The mill's collapse and the fire that followed ultimately claimed more than 120 lives and accounted for another 130 serious injuries. The victims in this story were mostly recent Irish immigrants and their children. The villains were the mill's owners, George Howe and David Nevins Sr., who had ignored their factory's structural frailties by cramming too many spindles, looms, and other heavy machines into the factory's upper floors. Neither they nor anyone else, however, ever compensated workers for their injuries.[40]

Similar scenes played out in other areas of manufacturing. In the winter of 1905, for example, a boiler explosion at the Grover Shoe Factory in Brockton, Massachusetts, sparked a massive fire that claimed the lives of 58 employees and left another 150 injured, many severely. In this case, there were no obviously culpable villains other than the boiler itself, which was old and unreliable. The factory's engineer no longer trusted it and had only fired it up because its replacement needed to be cleaned. When he did, the boiler exploded and rocketed through the structure's four floors before hitting the roof and spinning downward. The out-of-control boiler spewed burning coals across multiple floors and sparked a massive fire. Unlike the Pemberton disaster, however, the shoe factory fire led to tangible safety reforms. In its wake, Massachusetts passed the nation's most stringent regulations governing the condition and operation of factory boilers, and its boiler codes helped set national standards.[41]

The story of America's most unforgettable manufacturing disaster—the Triangle shirtwaist fire—has been told countless times but deserves revis-

iting: on March 25, 1911, in the heart of New York City's business district, smoke and flames poured out of the ten-story Asch Building's upper windows. There, the Triangle Company employed about five hundred people to make garments using modern, belt-driven sewing machines mounted on tables. Most of these employees were young Jewish and Italian women who had only recently immigrated to the country and who sat shoulder-to-shoulder sewing popular ladies blouses known as "shirtwaists." At about five o'clock in the evening, shortly before closing time, someone in the company's eighth-floor workplace apparently dropped a hot match or cigarette into a scrap bin and started the fire. Within minutes, the fire swept across the cloth-littered floor and spread upward, killing 146 garment workers, mostly young women. Nearly sixty of the victims leapt to their deaths from the upper floors, landing with a lasting thud in front of horrified onlookers. When firefighters made their way through the factory, they found fifty dead bodies piled against doorways and another thirty crammed into an elevator shaft. Unfortunately, the firefighters' ladders had only been able to reach the sixth floor, leaving the workers trapped.[42]

The Triangle fire received extensive media coverage, and in its wake the public learned the blunt truth about the shirtwaist factory: it was a tinderbox as well as a sweatshop. The bin of scraps where the fire likely started should have been emptied days if not weeks earlier, and a factory exit door had been locked to prevent employee theft. The structure that housed the factory, the Asch Building, was hazardous itself. Its fire hoses were old and rotten, and its water valves had rusted shut. Only one of the building's four elevators functioned properly, and the stairwells were too steep and narrow to allow for a headlong rush of terrified workers. To make matters worse, the building's outdoor fire escape was loose and rickety. Regrettably, safety rules were so lax that such negligence went largely unpunished. The two owners of the shirtwaist company, Isaac Harris and Max Blanck, eventually settled nearly two dozen wrongful-death claims for only seventy-five dollars per claim, leaving the general public infuriated. Within a year of the fire, progressive legislators in twelve states had set up publicly managed workers' compensation programs for injured employees and their dependents, regardless of fault, and several other states did so over the next few years. Cities across the country also strengthened their building safety codes.[43]

The Triangle fire is the archetype of labor tragedies, the one that always comes to mind when historians say "workplace disaster." Its story has all the

basic elements of a well-known narrative in the history of American workers: the plot unfolds in a crowded, unpleasant setting somewhere in the East,
or perhaps the Midwest, before states established comprehensive workers'
compensation programs and child labor laws. The victims were working-
class men and women struggling to make a living, including many recent
immigrants from European nations not well represented in America's social
and ethnic fabric. The villains were male employers committed to personal
gain at the expense of others, easily recognized and just as easily vilified. No
one ever described the Triangle's owners as fatherly figures who cared about
their employees, or even as hard-pressed entrepreneurs in an especially
competitive industry. The newspapers described them as cold-hearted cap-
italists with the blood of young women on their hands.

Our lingering fascination with this 1911 tragedy can be explained on several levels. To begin with, the story of so many young women meeting such
a cruel fate is compelling on its own. But the story also illuminates larger,
more abstract patterns and themes. The Triangle fire stands as a tall historic marker of injustice, a seemingly uniquely American tragedy. Its flames
reflected the oppression of the nation's recent immigrants and industrial
workers, and it exposed the heartlessness of unregulated capitalism. With
the villains paying such a small price for their sins, the fire also revealed
shortcomings of the American justice system. At the same time, however, it
demonstrated the nation's ability to learn from tragedies. Within its ashes,
historians have seen the origins of several progressive reforms, including the
constitutional amendment that finally granted women the right to vote. The
Triangle fire was a powerful, symbolic example of a workplace tragedy that
transformed lives and laws.[44]

Tragic events in the nation's service sector also proved transformative. Like
major transportation accidents, disasters in service-oriented businesses
almost always harmed more patrons than employees, sometimes by ratios
of fifty or even a hundred to one. But they tended to occur in large metropolitan areas, often in densely populated neighborhoods. These tragedies
were exceptionally shocking because they wrecked places that people generally assumed were safe. Everyone knew that steel mills and mines were dangerous places, and even steamboats and trains were widely associated with
ghastly disasters. By comparison, structures like theaters, hospitals, and
schools seemed like sanctuaries. The working people in these places certainly

Fighting the Triangle shirtwaist fire, New York City, March 25, 1911. This infamous workplace disaster killed 146 garment workers, mostly young women. Courtesy of Brown Brothers, Sterling, Pennsylvania

stood apart from workers in other areas of the economy. Though most provided services of some sort, many had clerical and administrative skills.
Some were highly trained professionals, such as entertainers, nurses, and
teachers. The incomes of these people varied widely. Among the group were
many salaried employees who enjoyed a relatively high standard of living,
but there were also many workers who earned low hourly wages. Interest-
ingly, employees in the service sector often survived disasters in their workplace because they knew the exact locations of escape routes, some of which
were only accessible to them.

Chicago's Iroquois Theater fire of 1903 highlights these patterns. Before
the fire destroyed it, the Iroquois was one of the nation's newest and finest
playhouses. It seated sixteen hundred people on three separate levels connected by marbled stairways that rose from the richly decorated grand foyer.
The place was designed to accommodate large stage productions, and thus
it had an elevated stage, an orchestra pit, and multistoried backstage dressing rooms.[45] On the afternoon of December 30, the theater's patrons were
enjoying a long-running musical comedy, *Mr. Bluebeard*, whose cast included
150 performers and dozens of local extras, when sparks from a high-intensity,
electrically powered floodlight accidentally set fire to nearby curtains and
stage props. Seeing the problem all too clearly, chorus girls, actors, and orchestra members started heading toward the backstage exits. Meanwhile,
theatergoers began fleeing for their lives. In the chaos that followed, several
entertainers backstage burst through two large doors used for stage props,
sending blasts of cold air into the theater, in turn feeding the fire. As the
blaze spread, patrons in the theater's upper level, many of whom were women
and children, found themselves trapped by the fear-stricken crowds that
filled the aisles and stairways below them. Within minutes, some six hundred people perished.[46]

The Iroquois fire may not seem like a workplace disaster. After all, it
killed only five people on the job—an actor, an usher, an aerialist, and two
aisle attendants. But the fire nonetheless created enormous problems for
everyone employed at the theater. The Iroquois's ushers, attendants, and
clerks all had to find new jobs after the disaster, as did its pit musicians and
stagehands. The actors, wardrobe workers, and other "players" involved in
the stage show could not just join another tour company, even in a big city
like Chicago. Consider the problems that chorus girls faced. On average, the
young women earned about twenty dollars a week (six hundred dollars in

2020), from which they paid for their own lodging and food. When the fire consumed their dressing rooms, many lost whatever money and jewelry they had (the women rarely left their valuables in hotel rooms). The weather created another problem. It was frigid at the time, and without money, simply staying warm proved challenging. The show itself could not go on. The company that produced it had lost elaborate stage props and costumes in the blaze and quickly went out of business. A few days after the fire, once police finished interviewing witnesses, the show's "footlight queens" boarded a train with the rest of the performers and headed back to New York City, where they were on their own. Like most people who survived the fire, they had terrifying memories that likely troubled them forever.[47]

Whatever one might call this tragedy, it left a lasting impact on public policies. When investigators concluded that the Iroquois had violated several local fire codes, cities ordered inspections of all theaters within their boundaries. The City of Chicago closed thirty-seven local theaters for more than two weeks to conduct inspections, and during that time the city strengthened building ordinances to make theaters safer. New ordinances required theaters to ensure all fire exits were easily seen and marked with signs, and prohibited theaters from locking exit doors during performances. They also required theater balconies to have their own separate entrances and exits, and set new standards governing the width of all theater hallways and aisles. Additionally, the new ordinances mandated that local theaters cover all stage scenery and decorations with nonflammable chemicals. New York City, Boston, and several other cities followed Chicago's lead. In Omaha, Nebraska, for example, public officials required all theaters to increase the number of exits and broaden many aisles, and that often involved removing entire rows of seats. In the golden age of vaudeville and silent films, these reforms benefited not only thousands of theatergoers but also thousands of stage actors, pit musicians, and all other occupational groups that theaters employed.[48]

A fire in 1929 at a medical clinic in Cleveland, Ohio, offers another vivid illustration of a service-sector catastrophe that spurred safety reforms. The Cleveland Clinic was a four-story medical diagnostic and observation center whose doctors also treated patients and performed minor operations. Similar to other advanced medical facilities of the time, the clinic had a maze of examination rooms, laboratories, and administrative offices as well as a large library and pharmacy. But the clinic also housed a special department where technicians used x-ray machines to photograph patients' internal or-

gans and bones. These modern devices represented a breakthrough in medical science, but they also posed an unacknowledged threat. The machines contained strips of film made with a highly flammable chemical compound that emitted poisonous gases under extreme heat, and the clinic kept more than a ton of this material in a basement-level storage area.[49] On May 15, 1929, a mechanic repairing a leaky steam pipe in the area noticed a yellowish cloud of dust hovering over an open bin that held the film, an indication that the film was decomposing. The warm steam in the air probably caused the problem, but the closeness of a bright incandescent lamp to the bin may also have been to blame. In any case, the mechanic doused the cloud with water, a reaction that triggered a crisis. Moments later, two explosive flashes sparked a fire that sent highly toxic vapors wafting up stairwells and elevator shafts. Doctors, nurses, and patients started collapsing in hallways and examination rooms. By the time firefighters arrived and took control of the situation, 123 of the 250 people in the clinic had died, including more than two dozen of the clinic's employees. Another 50 people were rushed to local hospitals.[50]

Like the Iroquois fire, the tragedy in the Cleveland medical clinic had nationwide implications. Medical facilities everywhere soon established new standards for labeling and storing nitrate film. Cities and states also began regulating the storage of the film and other hazardous materials. Just as importantly, insurance companies stopped indemnifying buildings that refused to safeguard such items, and film manufacturers began developing new types of film to replace their unstable nitrate-based products. There were still other implications to consider. The dangers that rescue workers faced while combating toxic fumes shaped firefighting practices. Cleveland and other cities began issuing gas masks to their fire departments. The realization that many burn victims at the clinic relied on taxis to get to local hospitals also convinced Cleveland as well as other cities to introduce their own municipal ambulance services. All of these changes boded well for medical workers and their patients.[51]

If these patterns of havoc and reform in the service sector are still unclear, consider the case of the New London school disaster, the deadliest school tragedy in American history. On March 18, 1937, in an oil-rich community of East Texas, a gas explosion ripped through a large public school in the late afternoon, just as the final bell of the day was about to ring. Nearly three hundred people—more than half of the school's occupants—died in a

Collapsed school in New London, Texas, March 18, 1937. Natural gas from nearby oil wells seeped into the school's basement and ignited, killing nearly three hundred students and sixteen teachers. Courtesy of New London Museum

heartbeat. Though most of the victims were children, sixteen of the school's employees also perished in the disaster, including a dozen young female teachers. Investigators ultimately blamed the disaster on the local school board's recent decision to "tap" into nearby pipelines of an oil company for the purpose of reducing its gas bills. Faulty connections to those lines had allowed colorless, odorless gas to seep into the school's basement area undetected. When an instructor in the school's basement-level woodshop turned on an electric sanding machine, she unknowingly ignited the gas-filled air and sparked a deadly chain reaction.[52]

Although courts dismissed lawsuits related to the 1937 school disaster, the tragedy nonetheless affected methods of managing hazards related to

oil and gas production. In its aftermath, Texas and other oil-producing states began requiring oil companies to mix natural gas with specific chemical compounds whose familiar rotten-egg aroma has helped in detecting gas leaks. The tragedy also prompted states to place new restrictions on engineering practices, including practices associated with the public's utilization of gas pipelines. Only two months after the tragedy, partly to prevent inexperienced individuals from tampering with pipelines, Texas passed the Engineering Practices Act, a trendsetting law that prohibited anyone from representing themselves as engineers unless licensed by a newly established state board. The act's basic purpose, its language read, was to "protect the public health, safety, and welfare."[53] In the Great Depression, when the federal government rolled out a host of new public assistance programs, these words delivered an encouraging message. But they also underscore a basic premise of this book: that businesses and public officials generally respond to large-scale catastrophes, and in ways that have registered as progress.

In the century before World War II, disasters destroyed large places of employment in all areas of the American economy, in different geographical regions, and during different periods. Again and again, these tragic events revealed a grim reality: that growing technical options in American life had sparked large-scale catastrophes. Yet these events also encouraged public leaders at the municipal, state, and federal levels of government to acknowledge the frightful underside of material growth. In their wake, even courts and legislators who had strenuously shielded their communities from outside influences often supported the expansion of government responsibility in public life. Businesses consequently found themselves under increasing pressure to address safety concerns and to abandon traditional ways of defending themselves against accusations of negligence. By the 1940s, an assortment of private practices and public policies had made countless work environments safer, which protected a wide variety of working people as well as anyone who entered their workplaces as consumers.

As the five case studies in this book show, industrial and technological developments continued to spark workplace disasters long after the war ended and in the process inspired new generations of policymakers to assume greater responsibility for public safety. Each of these studies focuses on a disaster that occurred far from the centers of the Industrial Revolution, in a region of the country that was largely undeveloped before the war began.

The first study investigates a massive explosion that rocked the waterfront community of Texas City in 1947, destroying a large new chemical plant and everything around it. The second explores a 1956 air disaster that sent two commercial airliners spiraling into Arizona's Grand Canyon, and the third focuses on a 1971 earthquake that wrecked two hospitals in Southern California. The fourth examines a deadly 1980 resort fire in Las Vegas, and the fifth a terrorist act that brought down a federal building in Oklahoma City in 1995. Taken together, these studies challenge common assumptions about what constitutes a workplace disaster and whom these tragedies affect. They also serve to expose the double-sided character of modern life.[54]

2

Chemical Plant Explodes in Texas City!

> It is a mistake, also, to suppose that it is the cheap, unskilled labor which
> suffers most from industrial accidents.
>
> Crystal Eastman, *Work-Accidents and the Law*

On April 16, 1947, just after eight o'clock in the morning, plumes of orange-hued smoke climbed from the French steamship *Grandcamp* docked in Texas City, a modern deepwater port along the western edges of Galveston Bay. From the nearby Monsanto Chemical Plant, the largest industrial enterprise in the area, employees watched the smoke waft across the waterfront, not realizing they were in grave danger. The ship's cargo included box loads of government-owned ammunition as well as nearly twenty-five hundred tons of fertilizer-grade ammonium nitrate (FGAN), a new chemical compound developed to increase food, feed, and fiber supplies. Though touted as a major technological achievement, the granular material contained ingredients used in the manufacturing of explosives, and it was highly unpredictable when exposed to heat. At the time, however, the deadly nature of the material was not well understood.

The large quantity of FGAN in Texas City reflected new directions in American foreign policy. In the late stages of World War II, the federal government developed a plan to send as much of the juiced-up fertilizer as possible to famine-stricken regions of the world, not just for humanitarian reasons but also to contain social unrest and stop the spread of communism. As part of the plan, the government transferred control of army ordinance plants in Nebraska and Iowa to private businesses that could manufacture and distribute the material. Those businesses sent the FGAN to Texas City

Firemen and dockworkers alongside the simmering *Grandcamp*, April 16, 1947. Moments later, the ship's cargo exploded, which triggered a calamity that killed 568 people and injured another 3,500. Associated Press

by rail in large paper bags that were often exposed to high heat. Warehouse workers unloaded and then stacked the bags high atop one another in poorly ventilated warehouses near oil refineries, oil storage tanks, and the big Monsanto plant, all of which contained highly flammable combustible materials. Such conditions set the scene for a disaster of unimaginable proportions.

At exactly 9:12 a.m., after dockworkers and sailors had loaded countless bags of FGAN aboard the *Grandcamp*, the ship exploded with a force so powerful that it destroyed everything around it. The destructive force knocked down people in Galveston, ten miles to the south, and shattered windows in Houston, forty miles to the north. Earth scientists detected it as far away as Denver, Colorado, an astounding thousand miles to the northwest. With strength similar to that of an atomic blast, the explosion created a giant wave of salt water that swept across the Texas City docks, knocking down whatever was still standing. And yet the moment of impact had not fully unfolded. As the water began to recede, a second explosion ripped through the

area, causing even more chaos and destruction. Several hours later, as fires raged, another ship near the docks exploded. The *High Flyer* contained not only a thousand tons of FGAN but also two thousand tons of sulfur, which made for an even more powerful blast. When the ship blew up, flaming debris flew into the heart of Texas City. One of the vessel's large propellers traveled nearly a mile inland.[1]

As of this writing, the Texas City disaster remains the worst industrial accident in American history. The event killed an estimated 568 people and injured 3,500 others. It decimated the big Monsanto plant along with large warehouses, grain elevators, loading cranes, and other structures on the waterfront. It also damaged nearly a thousand homes and many small businesses in the local community. The disaster created a sense of bedlam in the Gulf Coast area that lasted for days. Hundreds of emergency workers poured into Texas City, bringing with them blood plasma and water as well as firefighting equipment, food, gas masks, and other critical supplies. Journalists and camera crews converged on the city, too. Probably no one on the scene had ever witnessed such devastation, including veterans from World War II.

Compared to other industrial accidents of the past, the tragedy in Texas City has received relatively little scholarly attention. Its origins are still shrouded in mystery, and its larger historical significance remains unclear. This chapter explores a series of neglected questions about the disaster. Most obvious among these questions is how the disaster affected the large Monsanto plant and its workforce. At the end of World War II, the plant had stood proudly on the waterfront, adjacent to the boat slip where the *Grandcamp* exploded. Workers there had played a major role in the war effort, producing the basic ingredients of a new substance—synthetic rubber, on which military success hinged. In 1947 the plant employed some 650 people, about 450 of whom were on duty that fateful morning. In a matter of seconds, 154 employees died and another 200 suffered serious injuries. Roughly half of the injured remained hospitalized for more than two weeks. These details are part of a story as rich and meaningful as any workplace tragedy in the nation's history.[2]

The rise of the Monsanto plant and its workforce in Texas City has to be understood against the background of World War II and the significance of natural, crude rubber. In 1940, Japan's military expansion began to threaten America's access to large rubber plantations in Southeast Asia, where the

Black clouds over the Monsanto Chemical Plant in Texas City, April 16, 1947. The postwar disaster destroyed the recently built plant and killed 154 of its employees. Getty Images

nation obtained about 90 percent of its rubber supplies. The sticky, milky substance that workers "tapped" from the trees of those plantations had remarkable qualities that made rubber indispensable to industrialized nations. Rubber was elastic, watertight, and highly resistant to abrasion. When properly refined, it could be attached to wood and metal goods and even blended with textiles to make durable fabrics. In the nineteenth century, manufacturers had learned to make many useful products with rubber, from rubber bands and bicycle tires to sink valves, fire hoses, and electrical insulation. By the early twentieth century, crude rubber was one of America's most valuable imported goods. Without it, the nation's industrial machinery would have stopped.[3]

As American military advisors warned at the time, the nation's ability to defend itself depended on its sources of rubber. Artillery components, ma-

chine guns, armored vehicles, walkie-talkies, pontoons, tents, life preservers, mosquito nets, and many other military products contained the substance. It took about a ton of rubber to build a tank or heavy bomber and seventy-five tons to build a battleship. Soldiers carried more than thirty pounds of rubber in their shoes, helmets, and basic gear. Military medics treated wounded soldiers with tourniquets, surgical bandages, and other medical supplies made with rubber. Jeeps and ambulances built with rubber parts rushed the wounded to military hospitals, where they slept and recovered on rubber-lined mattresses.[4]

Faced with an impending crisis, business and political leaders found new ways of acquiring the raw material. Because rubber trees were indigenous to tropical regions of Latin America, the United States purchased large amounts of rubber from Brazil and its neighboring countries. At the same time, the federal government established the Rubber Reserve Company to ration and distribute the nation's rubber reserves. President Franklin D. Roosevelt even asked citizens to donate old tires, garden hoses, and other products made of rubber to the war effort. By 1941, however, after Japan signed an alliance with Nazi Germany, it was painfully clear that such strategies were no substitutes for the potential loss of Southeast Asian rubber. National leaders concluded that the future depended on developing a synthetic substitute for the raw, milky material.[5]

Chemists had long struggled to find practical substitutes for rubber, but the costs of developing synthetic rubber coupled with the inferior quality of the product made it impractical for industrial and military uses. In the late 1920s, researchers at IG Farben, a conglomeration of German corporations, took an important step in the development of the new product by blending organic compounds called butadiene and styrene that could be derived from crude oil and natural gas. After Adolf Hitler and his Nazi Party came to power in 1933, the company began producing large amounts of the material and thus set Germany on the road to war. By the time the nation invaded Poland in 1939, IG Farben had made more than fifty thousand tons of its synthetic rubber, known as Buna-S. Two years later, when Japan bombed Pearl Harbor, the manufacturer was making twice that amount annually. The mass production of Buna-S gave Germany a decisive advantage in the war, which added to the sense of crisis in the United States.[6]

Americans learned the basic recipe for making this material in the early 1930s, when IG Farben's chemists exchanged technical know-how with their

counterparts at Standard Oil of New Jersey. Business and political leaders, however, had relatively little interest in the product until Japan threatened the nation's natural rubber supplies. Even then, conservatives in Congress hesitated to finance a national synthetic rubber program on ideological grounds, viewing such programs as socialistic. Further complicating the nation's response to rubber shortages was the fact that butadiene could be derived from grain as well as oil. Farm-bloc politicians in the Midwest insisted the nation could produce synthetic rubber more cheaply and efficiently by tapping grain supplies rather than oil reserves. Oil companies and their allies in Congress disputed these arguments, however, and the federal government ultimately launched an all-out effort to mass-produce synthetic rubber based on German technology and processes.[7]

This important campaign rested on massive financial support from the Reconstruction Finance Corporation (RFC). Congress had established the RFC in 1932 to help businesses cope with the Great Depression. During the 1930s, the organization primarily made loans and purchased stock and other financial securities, but as the nation drifted toward war, Congress gave it new responsibilities. By the summer of 1941 the RFC set up its own subsidiaries to mobilize national resources, including the Rubber Reserve Company and the Defense Plant Corporation (DPC), both of which helped finance the construction and operation of new synthetic rubber factories. The agencies soon entered into negotiations with several firms that were capable of producing the material, including Monsanto.[8]

Founded in 1901 by John Queeny, a purchasing agent for a drug wholesaler in St. Louis, Missouri, Monsanto initially manufactured food and beverage additives such as saccharin, vanillin, and caffeine. World War I created new opportunities for Monsanto by cutting off imports from large German chemical companies, and by the 1920s the firm was making basic industrial products like sulfuric acid in conjunction with chemicals used in pharmaceuticals and dyes. In the 1930s, chemists at the company's central research laboratory in Dayton, Ohio, began experimenting with plastics that required working with styrene and other essential compounds used in producing synthetic rubber. Monsanto's work in this field convinced the federal government to help the firm build a new production plant capable of producing ten thousand tons of styrene a year, thus providing other firms with the raw materials they needed to make synthetic rubber.[9]

↓

After studying several locations, Monsanto executives decided to build the styrene plant on the site of an abandoned sugar refinery along the docks of Texas City. The location seemed perfect. The protective shores of Galveston Island and the nearby Bolivar Peninsular protected Texas City's coastline, and ships had long passed through the Gulf of Mexico and safely dropped anchor there. In the early twentieth century, when cotton was Texas's most valuable commodity, the federal government helped finance dredging operations in Galveston Bay, and some of the world's largest cargo ships began to move through the area. In the 1930s the annual number of vessels that arrived in Texas City rose from twelve hundred to four thousand. By the 1940s, Texas City had become one of the nation's most modern shipping facilities. Its piers, warehouses, cranes, conveyor systems, railroad terminals, and other structures made it a logical place to build any manufacturing plant.[10]

The city's excellent terminal facilities were not the only factor that attracted Monsanto. By the 1940s, Texas City had become an integral part of the southeastern Texas oil empire. Nearby Houston boasted the greatest concentration of oil refineries in the world, and those companies could provide Monsanto with the raw materials it needed to make styrene, such as benzene and ethylene. A few refineries had recently sprung up in Texas City, too, including the giant, vertically integrated Pan American Petroleum and Transport Company, which owned one of the world's largest oil tanker fleets. Petroleum products already left Texas City on a daily basis. Railroads and oil pipelines stretched from the rich East Texas oil fields straight to the city's docks.[11]

There were still other reasons to invest in Texas. The price of land along the state's coastline was relatively low, and construction costs were low, too. Labor unions in the state were relatively weak, and state officials had long privileged the interests of management over those of labor. Texas oil barons had enormous clout in state politics, and the state's regulatory environment was therefore friendly to petroleum-related developments. Additionally, some of the machines in the old sugar factory that Monsanto purchased were still operative, including large oil boilers, electrical generators, and switchboard equipment.[12]

Engineers, draftsmen, and technicians began designing the plant in late 1941, just days before Japan bombed Pearl Harbor. H. K. Eckert, one of the plant's original managers, recalled Monsanto recruiting these men, suggest-

ing the company appealed to their sense of pride and patriotism: "Our company sent out word to all its plants for men who would shoulder the burden of long hours, hard work, and disruption of their normal lives to build a plant which would produce the material that was so vital to their nation's welfare," Eckert explained. "The response was beyond all expectations." Eckert painted a picture of dedicated workers doing all they could for the company: "Men of all ages and positions volunteered to help. They came willingly, anxious to contribute what they could toward the success of this program."[13] In 1942, when the plant was still under construction, Monsanto advertised its remaining job opportunities in Texas City and received applications from people far and wide. Looking back on that year, Eckert recalled that people flocked to the plant in search of work, again picturing workers as a noble lot: "They came from the surrounding country to offer their services, eager to take part in this work so vital to their country."[14] On closer inspection, the flood of job seekers largely reflected the lingering effects of the Great Depression, which hit the coastal area especially hard. When the chemical plant advertised its jobs, many people there were in desperate straits.

The construction of the plant testified to the ingenuity and stamina of many people that Monsanto employed. That task required twenty-five hundred workers from various trades and involved dredging and draining a large part of Galveston Bay before construction even began. Workers erected several technologically advanced structures on the site, including an advanced research laboratory as well as large chemical distillation and purification units. They also built large warehouses, storage tanks, and a variety of other structures. Overhead pipelines and service roads connected these disparate parts of the enterprise, creating one of the nation's most technologically complex business environments. With his characteristic flair for embellishment, Eckert said the plant rose from the ground "like a magnificent flower." "It was built according to the plans evolved by the minds of men," the manager recalled nostalgically. "It was a product of their minds and as much a part of them as any member of their body."[15] To Eckert and other company men, this was progress, plain and simple.

Work routines within the plant revealed a setting different from the one Eckert described. In many ways, they resembled manufacturing operations in nearby oil refineries, where employees heated, fractured, and compressed crude oil and its by-products. Producing styrene required employees to work with several petroleum by-products. It involved jobs like starting and stop-

ping pumps and compressors, setting and changing valves and gauges, and testing and analyzing chemicals. It also meant monitoring temperatures within processing units ranging from below zero to more than a thousand degrees Fahrenheit. The finished product was a colorless, oily substance that evaporated easily and was usually shipped in large metal tanks or cardboard drums.

By the time the first containers of the substance left Texas City in March 1943, the federal government had contracted with Monsanto to increase production of styrene to fifty thousand tons a year. To accomplish that goal, Monsanto had construction workers dredge more sand from Galveston Bay for the purpose of extending the margins of its waterfront property. On its newly raised soil, the company built new processing units and storage facilities. But increasing production levels also meant improving manufacturing techniques, including processes used to treat hydrogen and other chemical elements during the production process. Although a hurricane swept through the bay area in the summer of 1943 and slowed the plant's expansion, in early 1944, Monsanto was making even more styrene in Texas City than the government expected. By the end of the war, the Texas plant had produced nearly a quarter of all the styrene in the government's synthetic rubber program and thus made a major contribution to the nation's war effort.[16]

The plant's production of styrene dropped markedly after the war, but synthetic rubber remained valuable, and Monsanto continued to manufacture it in large quantities. In 1946 the company purchased its Texas City plant from the federal government and began selling its products to manufacturers in the emerging plastics market. At the same time, it also began producing a new synthetic polymer—polystyrene—that was used to "harden" products made with plastic and rubber. That strategy required Monsanto to invest in polymerization units in order to break styrene down into separate components. In early 1947, after Monsanto built yet more chemical processing units on its property, the company added a separate "Texas Division" to its organizational structure, a clear indication of its recent success.[17]

On the eve of the catastrophe, then, Monsanto's plant loomed large in the bay area. Its steel-reinforced structures, pipelines, and heavy machinery stretched across a rectangular tract of land covering forty-three acres. From across the bay, people could not help but notice the plant's two chemical distillation towers, which were nearly two hundred feet tall and had large control rooms, pumping stations, and storage tanks at their bases. Other

parts of the plant stood out, too, including a five-storied polystyrene processing building and sixty-foot-tall ethylene purification units. The plant's
laboratory, instrument shop, propane tanks, and other structures were
harder to see but no less important to the production process. Though Monsanto had chemical plants in other parts of the country, its enterprise in
Texas City was unusually impressive. But the plant nonetheless posed a
threat to the waterfront community. Unfortunately, the threat was largely
invisible.[18]

In early 1947, Monsanto's Texas City plant employed about 650 people, a
large majority of whom had been there since the day it opened. The plant
divided these employees into two general categories—salaried employees
and hourly wage earners—each of which had its own subcategories. More
than 150 of the plant's employees, or nearly a quarter of its workforce, earned
monthly salaries. Most of these employees had extensive training and work
experience, and their jobs demanded technical skill and creativity. Those who
earned the highest salaries were chemists who understood the processes involved in making styrene and polystyrene, or engineers involved in building
and expanding the plant.[19]

All of these salaried employees were white males, a blunt fact that reflected both custom and law in the labor market. Across the country, and
especially in the nation's Sunbelt states, jobs involving authority and high
pay were almost always reserved for Caucasian men. Before the civil rights
laws of the 1960s took effect, businesses blatantly discriminated against
women and minorities. Public ordinances and cultural practices kept cities
and industries racially segregated. Attitudes and customs discouraged if not
barred women and minorities from attending institutions where chemists
and engineers learned their job skills. Trade schools and labor unions embraced discriminatory practices, too, which made the challenge of acquiring
skills all the more difficult. In these years, gender and racial equality was not
part of Monsanto's hiring strategy, nor was the matter of much concern to
other businesses along the Texas coast.

The background of Paul Jacobs, who perished in the 1947 disaster, offers
more insight into the lives of these higher-salaried employees. Jacobs was
one of the plant's leading chemists. He began working for Monsanto in 1936,
only twelve days after graduating from Wittenberg College in Springfield,
Ohio, where he majored in chemistry. Jacobs was one of the college's most

gifted students. He was a prominent member of the institution's chemistry and language honor societies, won prizes in annual academic contests, and graduated in the top 10 percent of his class. Though he could have entered a widely respected graduate program at the University of Wisconsin, Jacobs instead accepted a job at Monsanto's research laboratory in Dayton, Ohio. When the war began, Monsanto sent Jacobs to Texas City to help produce styrene, with a monthly salary of $455, or $5,460 a year (worth about $80,000 in 2020).[20] At a time when a gallon of gasoline cost less than twenty cents and a loaf of bread could be had for a dime, that was a good salary. It was more than enough to buy a home in Texas City.

Bernard Merriam was another highly skilled employee who died in the disaster. Merriam had graduated from Harvard University in 1936 with a degree in engineering and worked at Monsanto's headquarters in St. Louis before moving to Texas City, where he helped design the company's styrene plant. According to plant manager Eckert, Merriam played a major role in expanding the plant during the war years and was one of its best engineers. "Mr. Merriam had proven his ability and was advancing to a high executive position in the Texas Division," Eckert reported after the disaster. Like Jacobs, Merriam was married with three small children when the disaster took his life. He was one of the plant's highest-paid employees, earning a monthly salary of $640 a month, or $7,680 a year ($115,000 in 2020). In other words, Merriam was relatively well paid, even compared to other skilled workers at the plant.[21]

About fifty of the plant's salaried employees were white-collar workers, such as accountants, secretaries, and telephone operators. Many of those individuals had considerable technical and administrative skills, and often made independent judgments in the execution of their tasks. Although some had no more than a high school education, many were mathematically adept and had business school training. Almost all knew how to operate a variety of office machines. The responsibilities of these employees included tasks such as preparing monthly maintenance cost reports, maintaining payroll records, and assisting with salary payroll distribution. Some compiled daily and monthly inventories, prepared and analyzed shipping records, or wrote letters of transmittal. Others performed general clerical and stenographic duties as required. Although men held two-thirds of the plant's salaried positions, women held about three-quarters of these white-collar jobs. Like the chemists and engineers, they too were white.[22]

The background of Marie Dalrymple, a senior clerk at the plant, provides a glimpse into the world of this occupational group. Born in Longview, Texas, in the northeastern part of the state, Dalrymple graduated from high school in 1923 and spent a year at two different small colleges in Texas before getting married and taking a full-time job at Longview's County Clerk's Office. Over the next ten years, she held clerical jobs at a local insurance company, clothing store, and pipeline company before moving to Texas City, by which time she and her husband had divorced. In the Gulf Coast area, Dalrymple found secretarial work at an insurance company and a construction firm before landing a similar job at the Monsanto plant in 1944. The job only paid seventy-five cents an hour (about eleven dollars in 2020), leaving Dalrymple among the company's lowest-paid employees. At her supervisor's suggestion, however, the woman enrolled in a local business school to take courses in accounting, and Monsanto then transferred her to its accounting department, where she earned a monthly salary of $255, or $3,060 a year ($46,000 in 2020). The new job came with new clerical and bookkeeping responsibilities, such as reviewing invoices and keeping track of company expenditures. Dalrymple survived the 1947 disaster with only minimum injuries, but her second husband, who worked at the nearby Republic Oil Company, perished.[23]

Nearly five hundred people worked at the plant for hourly wages, including both skilled blue-collar workers and comparatively unskilled general laborers. The former group included craftsmen like carpenters, painters, boilermakers, pipefitters, and electrical workers. Though some of these workers held high school degrees and attended trade schools, many acquired their skills in the plant's apprenticeship programs. By the time they became journeymen, they knew how to use a variety of modern tools, like lathes, drills, and measuring instruments. Some also learned to install and repair complex machinery, such as water pumps, electrical generators, boilers, drying ovens, and switchboards. These workers were also white males.[24]

The work experiences of Earl Grice illustrate these patterns within the blue-collar force. Grice had relatively little formal education and no special skills when Monsanto hired him in 1943 as a pipefitter's "helper." In that job, he earned an hourly wage of ninety-two cents (about fourteen dollars in 2020). Grice's job involved assisting skilled workers in their daily tasks, keeping them supplied with materials and tools, and cleaning their work areas and equipment. Once Grice grew familiar with a pipefitter's tools and tasks

(about six months later), the plant reclassified him as an "apprentice" and raised his hourly pay to $1.07. He remained in that position for more than a year, during which time he received five merit raises before finally earning the title of pipefitter, when his pay rose to $1.39 per hour. By then, Grice could fabricate, assemble, install, dismantle, test, and repair piping made of metal, tile, glass, or porcelain. Like other skilled men in his trade, he used saws, grinders, benders, and welding devices as well as hand tools like tapping machines, wrenches, and rigging equipment. Had the disaster not taken his life, Grice would no doubt have been promoted to the rank of "leadman," supervising the work of pipefitters. By 1947, he had already received two more merit increases and was earning $1.79 an hour. In effect, Grice had nearly doubled his hourly wage in only four years. By most any yardstick, that should have been a story of success.[25]

General laborers were the plant's least skilled and lowest-paid workers. They held the most physically strenuous and monotonous jobs and were often employed on a seasonal or temporary basis. Many of these laborers worked outdoors on construction projects, digging excavations, dismantling scaffolding, or loading or unloading trucks. Like unskilled "helpers" within the plant, they often assisted skilled workers, mixing mortar for bricklayers, for example, or carrying tools and supplies for carpenters and painters. A few worked as porters and janitors cleaning the interiors of structures, sterilizing rubber boots and apparel, washing company-owned cars, and performing other such duties as required. A few others were security guards whose jobs involved keeping unauthorized persons from entering the plant or reporting unusual or suspicious activity in the area. A large number of these employees had no more than a few years of education, and almost all were men. At least fifty of the laborers were African American, and a few others were Mexican or Mexican American. To make ends meet, many of these workers held other jobs in the area while employed by Monsanto. Some worked part time loading and unloading ships' cargo for local stevedore companies, and others occasionally worked for small businesses in the area.[26]

These lowest-paid employees deserve to be viewed as more than just common laborers. Their work experiences can only be understood against the backdrop of race and racism in the area. The case of Jim Goff, one of the black workers at the plant, illustrates the point. Born in Brenham, Texas, about 120 miles northwest of Texas City, Goff had only a fifth-grade education when Monsanto hired him as a day laborer in 1946. He had an impres-

sive military service record, however. During the war, Goff served as a truck driver for the US Army and participated in campaigns from north Africa to the beaches of Normandy and the densely forested Ardennes region of Belgium. By the end of the war, the army awarded Goff the prestigious European, African, and Middle Eastern (EAME) Ribbon along with five bronze stars and a Good Conduct Medal. Despite this remarkable record, Goff could apparently find no better job after the war than the one he held at the chemical plant, where he engaged in mundane tasks like moving and piling plastic barrels within the old sugar refinery building. To supplement his hourly wage of eighty-eight cents, Goff occasionally took part-time jobs at a local grocery store, a beauty shop, and other small businesses in Texas City. And yet Goff was among the fortunate few Monsanto employees who survived the disaster with only minor lacerations and abrasions.[27]

About eighty of Monsanto's employees had supervisory duties and thus spent much of their time ensuring that their subordinates completed their daily tasks according to job specifications and deadlines. This often entailed training employees as well as making work schedules and handling grievances in the workplace. It also meant writing employee evaluations, interviewing job applicants, and recommending promotions for employees under their watch. The supervisors, almost all of whom were white males, typically received individually negotiated salaries, but they were well below the circle of high-level managers who oversaw entire departments within the plant.[28]

The duties of Paul Axe, the superintendent of loading, offer a clearer view of the supervisors' world and the organization of their workplace. Hired in 1943, when the plant was still under construction, Axe supervised more than a dozen employees involved in the handling of liquid materials that came in and out of the plant in cans, drums, and tanks. In doing so, he had to stay in close communication with the plant's Purchasing Department to track sales and purchases of liquids and to make sure there were adequate storage facilities for the material. He also had to be in close contact with the Traffic Department regarding trucks, barges, and trains entering and leaving the waterfront area, and work with the Maintenance Department to keep equipment and storage tanks in operating order. Furthermore, Axe worked with the plant's Control Laboratory to help test materials his employees handled, and he communicated with other departments to understand production schedules and storage inventory. Axe earned a bachelor of science degree in mechanical engineering from Texas A&M, and in 1947 he earned a monthly

salary of $410, or $4,920 a year ($58,000 in 2020). Although he survived the disaster, Axe suffered serious injuries to his hands, back, and legs.[29]

The vaporization of the *Grandcamp* blew out the walls of businesses along the waterfront, but the blast also sent large pieces of red-hot steel flying through the area like bunker-busting bombs. Pieces of the ship weighing up to two thousand pounds traveled more than a quarter mile from the docks and burrowed more than six feet below the ground. The *Grandcamp*'s cargo included many balls of sisal twine that ignited and spiraled through the air like medieval thermal weapons. Only seconds after the *Grandcamp* exploded, a fifteen-foot wave swept across the waterfront, causing more chaos and destruction. By the time the water receded, most businesses near the docks had suffered major property damage. Whatever was still standing suffered another mighty blow several hours later, when the *High Flyer* exploded. The condition of the Texas City Terminal Railway Company, whose workers handled shipments of ammonium nitrate at the port, reflected the extent of the damage. The company lost several of its warehouses and auxiliary buildings together with a large grain elevator and all its contents. The refineries of Republic Oil, Stone Oil, and the Humble Pipeline Company took knockout blows, too.[30]

Monsanto's chemical plant suffered heavier losses than all of these companies combined. The initial blast from the dock obliterated the plant's main warehouses and severely damaged its research laboratory, main office buildings, and large pump house. It also hollowed out the old sugar refinery building and a group of large brick and steel structures around it, and it almost felled the plant's twin distillation towers. Fires in the plant fed on benzene, ethylene, and other combustible liquids that spilled from ruptured pipelines, storage tanks, and purification units, creating chain reactions that compounded the disaster's destructive effects. Fires inside and outside the plant raged for two days, not only gutting more structures but also delaying rescue and relief efforts. The giant wave that washed across the area exacted its own toll on the company and its employees. It left a thick layer of oil and styrene over whatever remained of the plant's machinery and equipment, and drowned an untold number of people.[31]

Employees who survived the disaster remembered these terrifying moments all too well, and their recollections help to explain both the devastating nature of the catastrophe and the various ways people reacted to it. What

A fireboat douses water on Monsanto's hollowed-out styrene plant, April 16, 1947. During World War II, Monsanto converted an old sugar refinery on the Texas Coast into a modern workplace that produced the basic ingredients of synthetic rubber. Houston Press Photographer, RGD0005-f6320, Houston Public Library, Houston Metropolitan Research Center

happened to Charles Yaos, an accounting clerk in the Shipping Department, mirrored the general pattern. Yaos was on the ground floor of the dock warehouse telling three other employees where to move heavy drums of plastic. When he learned that a small fire had started on the *Grandcamp*, docked just outside the warehouse, the clerk was concerned but not alarmed. "We had been warned of the possibility of fumes from the fire on board the *Grandcamp* and had been told that if the wind changed it would be best for us to get out of the area," he later told investigators. "We had not been told of any possibility of an explosion." But the blast from the ship almost killed Yaos. "I was blown straight up in the air," he recalled. "Water came over me and I attempted to swim and then the water came down and I found myself on a pile of rock, brick and other debris, buried up to my waist with rock and brick." Covered with oil and blood and with serious injuries to his chest and legs, Yaos staggered out of the warehouse toward the plant's entrance, where

someone helped him into an automobile and rushed him to a nearby hospital. The clerk remained hospitalized for ten weeks. Yaos would return to the hospital more than a dozen times to have pieces of wood and other debris removed from his body.[32]

One of the plant's day laborers, Willie Davis, had a similar experience. Davis was loading styrene into railroad tank cars outside the warehouse when the ship exploded. "It felt like I was shot in the back with a shotgun," he later recalled. The explosion thrust Davis under the car and almost blinded him. "I peeped out just enough to see that the tank car wheels were a foot off the tracks. Then came the tidal wave." Davis tried in vain to hold on to the car. "The water took me, train and everything else with it. My face, body and head felt like they were going through one of those old-fashion washing machine wringers." Thinking he had been swept out to sea, Davis expected to drown. But he soon found himself on solid ground, covered with oil and tar and practically naked. The terror continued. With fires breaking out all around him, Davis struggled to his feet. "I found a small gully running under the fence with a body wedged under it. I pushed the body out with my feet and crawled through." He then ran toward town until he spotted a passing bus, which he jumped onto before falling unconscious. "I came to on a mattress on the floor of St. Mary's [Hospital] and remained semi-conscious for several days."[33]

Vernon Bruggman, who worked in the plant's receiving department, had his own story to tell. Like several of his coworkers, Bruggman heard about the fire on the *Grandcamp* and wanted to see it firsthand. His supervisor, however, warned employees that water was boiling around the ship and told them to remain indoors. A few minutes later, Bruggman almost lost his life. The explosion threw him to the ground, where he cracked his head against concrete. "My head bounced on the concrete like a rubber ball," he recalled. "The noise was unbelievable. I stood up, and chips and chunks of concrete hit my head from the floor above." As he was checking for broken bones, a huge wave of salt water crashed through blown-out windows, and then the second explosion occurred. The twin blows flung him from the building and left him submerged in knee-high water. Bruggman again managed to stand. "Dazed, I had blood running in my eyes and couldn't see too well." Noticing that nearby buildings were on fire, he decided to run for his life. "I didn't want to burn alive, and I headed toward my car." But the plant's parking lot had turned into a smoking junkyard, so Bruggman stumbled toward town.

Along the way, he witnessed a few unforgettable scenes. "A young boy ran past me, crying and screaming, his right arm almost blown off, dangling as he ran." Moments later, the boy "stepped on a live electrical wire and was killed." The shaken Bruggman eventually made it to a hospital, where he spent the next nine days. Like so many survivors, he had to return to the hospital several times to have his wounds treated.[34]

Mary Hunter, a secretary at the plant, was sharing a morning snack with a coworker when the *Grandcamp* started smoldering. Though Hunter saw smoke and fumes in the sky, she never realized the danger it posed. "Lorna and I were watching the fire and then decided to stop and have a sweet roll. We broke one in two and that's when everything went black." The initial explosion threw Hunter to the floor, where she remained unconscious for at least a few minutes. When she came to, Hunter realized she was seriously injured and alone in a bombed-out building. "I was on the ground floor of a three-story building and could look up and see the sky." Hunter called for help repeatedly before someone found her and carried her to a smashed-up car outside the plant, leaving her there. "I was bleeding and thought I'd lay there and bleed to death if I didn't do something." Someone else, perhaps a rescue worker, eventually found Hunter and rushed her to a local doctor's office. The explosion left her with severed nerves in her arms and deep cuts from the top of her head to her ankles.[35]

Jewell Robinson was on the ground floor of Monsanto's distillation unit taking paper charts from a machine when her world changed profoundly. The initial blast felled Robinson and left her covered with debris. "When the explosion occurred, something knocked me down and the ceiling fell down on me," she later explained. In the smoke-filled room, the young clerk looked about frantically for help and quickly spotted a male coworker. "I could barely tell who he was," she said. When Robinson called to him, the man rushed to her side, helped her up, and told her to run. The clerk ran about a hundred yards and jumped down a bank when the second explosion occurred. At that moment, she was too petrified to move. "I don't know how long I stood there; I heard someone calling me from behind, and it happened to be one of my neighbors." The neighbor, a man who also worked at the plant, lifted Robinson to the top of a fence that enclosed the plant, and another man helped her down. From there, Robinson limped to a nearby drugstore, where her husband eventually found her. With deep lacerations to the head and injuries to her lower back and legs, she then made her way to a local hospital.[36]

Many survivors, including Jewell Robinson, lived in the working-class neighborhood that bordered the waterfront area and thus lost much of their personal property that tragic day. The explosion left Robinson's home in ruins. It lifted her house off its foundation, pushed out several walls, and brought down large parts of the roof. It wrecked her home's plumbing system and gas lines, and obliterated windows, doors, and furniture. Robinson could not even find many of her kitchen appliances after the disaster, or her dishes, glasses, pots, and pans. To make matters worse, the explosion badly damaged a trailer on her property that she rented to her brother-in-law. Insurance agents later estimated that the catastrophe reduced the total value of the Robinsons' property by more than 75 percent.[37]

The story of Robert Bailey, a laborer who lived in the working-class neighborhood, also reveals the disaster's impact on Monsanto's employees and their homes. Bailey was standing on a ladder chipping paint from a large storage tank when the explosion hurled him through the air and left him unconscious. When he finally regained his wits, Bailey staggered into town and found someone to drive him to a medical clinic. Over the next two weeks, doctors at four different hospitals treated Bailey for a separated collarbone and broken ribs as well as injuries to his eyes and ears. Like his neighbors, Bailey also lost much of his personal property in the disaster. The explosion ruined his home's roof, walls, floors, closets, windows, and doors. It ruined his home's furniture and even the clothes in his closets. It also killed more than two hundred of Bailey's chickens that he and his wife had raised in their backyard.[38]

As news of the disaster spread, emergency response teams from Houston, San Antonio, and several nearby smaller cities rushed to the disaster site. Firemen fought fires there while police officers directed evacuation efforts, controlled traffic, and tried to calm terrified citizens. The nation's largest and most well-organized humanitarian organizations—the American Red Cross and the Salvation Army—sent dozens of volunteers to the area, where they set up relief stations, distributed blankets and clothing, and helped get the injured to hospitals. Doctors, nurses, and morticians from across the region rushed to Texas City, too, and helped staff local hospitals and medical clinics. Even Boy Scouts came to the city to help with tasks like sweeping glass from sidewalks and distributing supplies to relief stations.[39]

The federal government played the most crucial role in relief efforts.

When President Harry Truman received news of the disaster, he immediately issued an executive order instructing all federal agencies to do whatever they could to assist disaster victims. Truman telegraphed Texas City Mayor J. C. Trahan to relay this message and to express his concern for the people: "I have asked every Government agency to cooperate in relief activities. My heart and the heart of the nation go out in deepest sympathy to the suffering people of Texas City."[40] By standards of the time, the federal response to the disaster was impressive. The army rushed more than seventy of its doctors and nurses to Texas City, along with five thousand units of blood plasma, six thousand blankets, and one thousand cots. It also set up two fully staffed and well-supplied makeshift kitchens to feed the hungry and brought in bulldozers, tractors, cranes, and dump trucks to help with rescue and recovery operations.[41] The navy sent planes and trucks filled with blankets, stretchers, and medical supplies to the city as well as two dozen of its doctors and nurses.[42] The US Army Corps of Engineers coordinated these emergency operations from its Galveston District and provided a range of other vital services. Its corpsmen helped to reestablish telephone lines, for example, and to dredge submerged debris from the waterfront. They also transported disaster victims to nearby morgues and hospitals. In the harbor itself, the US Coast Guard helped to locate and transport debris and dead bodies from the water, and to keep ships away from the area.[43]

The media's response to the event was immediate and wide ranging. Galveston's main radio station broadcast an emergency bulletin only three minutes after the first explosion. There was so much interest on the airways that the Federal Communications Commission made additional airwaves available for amateur radio operators, hoping the move would help the general public assist people in the devastated area. Local and regional newspapers had reporters on the scene within hours, interviewing anyone they could find and photographing everything. By the end of the day, the nation's leading newspapers had their own reporters in Texas City, looking to understand and explain the tragedy to their own readers.[44] The intensive news coverage made the tragedy seem all the more real, tangible, and visible. For days after the event, the front pages of major newspapers featured disturbing images of devastation and chaos. US Congressman John Connally of Texas underscored the salience of this news coverage while informing the House of Representatives of the disaster: "The press is filled with accounts of one of the great holocausts ever visited upon the United States." Front-page pictures of

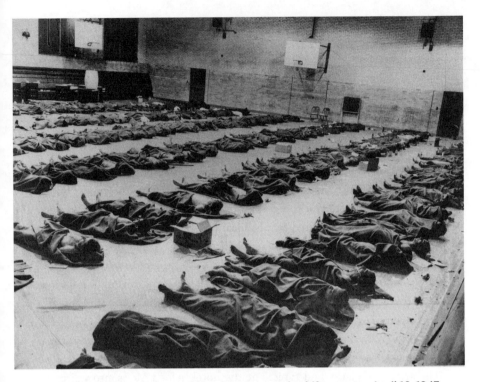

Texas City disaster victims scattered across a makeshift morgue, April 16, 1947. The number of Monsanto employees in this photo is unclear, but the company's workers accounted for more than a quarter of all fatalities. Houston Press Photographer, RGD0005-f6320, Houston Public Library, Houston Metropolitan Research Center

smoke-filled skies, burning structures, demolished cars, and dead bodies sparked passionate reactions.[45]

Monsanto reacted to news of the disaster quickly. Several of the company's top executives flew immediately to Texas City, including its founder's son, Chairman of the Board Edgar Queeny, and President William Rand, both of whom were on the East Coast at the time. Before landing, the company's private plane circled above the still burning waterfront area, which reminded Queeny of recent film footage depicting atomic tests in the Pacific. The chairman described the scene a few days later in a graphic letter to the company's stockholders: "Fires still raged in both Humble and Republic Oil Companies' nearby tanks, raising a huge Bikini-like pillar pall 3,000 feet into an otherwise clear sky." "A mile-high Vesuvian wick joined the titanic smoke

columns arising from the near-by storage tanks of oil refineries, forming a pitchy umbra over the racked and grief-laden community."[46] Other Monsanto representatives set up a command post in the lobby of a Galveston hotel to collect information about the causes of the disaster and its impact on the company's employees. A few of the representatives took taxis toward Texas City in hopes of surveying the plant but could not get close enough to do so. Others visited local morgues and hospitals and started making lists of the dead and surviving employees.[47]

Monsanto struggled for days to dispel the notion that an accident within its Texas City plant had somehow caused the disaster. As Chairman Queeny noted in his letter to stockholders, some of the initial news reports coming out of Texas City suggested that Monsanto either produced ammonium nitrate or had used it in its manufacturing processes. "Some erroneous statements were circulated," as Queeny put it.[48] To set the record straight, Monsanto sent personnel from its Department of Industrial and Public Relations to Galveston, where they voluntarily discussed the disaster with reporters at the company's hotel command post. "We made every effort to be of service to them in the hope that we would be able to steer their stories as far as possible in the right direction," one of the representatives later reported. The arrival of clerks and typists at the hotel helped Monsanto write and distribute its own press releases. What happened in Texas City was nonetheless a public relations nightmare. Despite the denials made by company officials, the volatility of Monsanto's raw materials clearly contributed to the disaster. True, the plant did not manufacture ammonium nitrate, nor did it use or store the material. But it did produce flammable products like benzene, styrene, propane, and other raw materials that burst into flames when ignited.[49]

As the company's task force in Galveston soon learned, more than 150 of the 450 employees on duty at the plant were dead or missing, including some of the top supervisors and best technicians. The plant's chief engineer, technical director, maintenance superintendent, and head safety inspector had all died. So, too, had sixteen of seventeen chemists as well as many of their key assistants. The disaster killed almost all of the men working in the plant's warehouse and a large crew of construction workers employed by outside contractors. News reports later claimed that the disaster killed 225 people within the plant's boundaries, which accounted for more than a third of all fatalities on the waterfront.[50]

The chairman and president stayed in the area long enough to watch bull-dozers start clearing debris from the waterfront, an operation that turned up more dead bodies. Seeing emergency workers placing the charred bodies on stretchers left a deep impression on the men. "It was heart-breaking and gruesome; memories of it will haunt us forever," Queeny said of the experi-ence. The men found it equally difficult while visiting local hospitals to see so many employees struggling to survive. "As we visited with our own stricken and saw the conditions of others, it was impossible for us to contain emo-tions," Queeny explained.[51]

As rescue and relief efforts proceeded, Monsanto announced a multi-pronged approach to assist its employees and their families. The plan tapped the company's life insurance policies in order to pay widows or beneficia-ries of its deceased employees lump-sum payments from $6,000 to $20,000 ($70,000 to $235,000 in 2020), depending on their loved ones' former in-comes. Monsanto also had insurance coverage through Texas's workers' com-pensation laws, which provided an additional $7,200 to the bereaved. The company's assistance extended beyond its insurance policies and legal liabil-ities. Monsanto's Board of Directors established a $500,000 (nearly $6 mil-lion in 2020) emergency relief fund to cover property losses resulting from the disaster and to help families relocate to homes outside the area. The fund also provided injured workers with their regular incomes while they convalesced and paid hospital bills that exceeded costs covered by employee health insurance plans.[52]

Monsanto also reassured its employees that it intended to rebuild its Texas City plant and rehire all workers who wished to return to their jobs. President Rand encouraged survivors who were then capable of working to report for duty at once. Some could clear debris from the plant or clean whatever equipment and machines were still operable. Others could help re-place lost personnel files and business contracts or estimate financial losses resulting from the disaster. At roughly the same time, Chairman Queeny assured stockholders that Monsanto had insured its Texas City plant for twenty-five million dollars (nearly three hundred million dollars in 2020), or enough to rebuild damaged structures and replace lost machinery and equipment. "Of course the loss of such an important unit will have an ad-verse effect on future profits," Queeny admitted, "[but] we will build again at Texas City."[53]

The rebuilding process was a multifaceted, herculean task. It involved

contracting companies to remove tons of twisted steel from the waterfront and decontaminate the area, for example. It also involved renting temporary offices in downtown Texas City to reestablish Monsanto's personnel, accounting, and purchasing departments. With an eye toward expanding its chemical operations along the docks, Monsanto purchased a large tract of land from the Texas City Railway Company where it planned to produce vinyl chloride, a recently developed organic compound used to manufacture industrial products. While pursuing these and related strategies, the company also focused on reconstituting its large and diverse workforce. That meant advertising jobs, interviewing applicants, preparing personnel papers, and arranging new work schedules. It also meant familiarizing staff with new work environments and convincing them that those places were safe. Monsanto's ability to bring back employees who survived the disaster spoke to how those individuals viewed the company. In 1948, when Monsanto finally resumed production of styrene in Texas City, it had rehired 95 percent of its surviving employees.[54]

The employees who never returned to the plant fell into three general groups. The first group encompassed workers whose injuries were so serious that they never went back to work. Hannah Holloway, a dishwasher in the plant's cafeteria, offers a representative example. At the time of the explosion, Holloway was carrying out her daily work routines when the cafeteria's ceiling suddenly crashed down upon her. Boiling liquids from the nearby kitchen stove then splashed across her face, chest, and right arm, causing further injuries. Holloway remained hospitalized for several months after the disaster, during which time surgeons had to amputate her right leg. One of the doctors who performed the surgery said Holloway would never fully recover from her injuries. "She will hereafter continue to suffer physical and mental pain throughout her life," he wrote in a medical report. Holloway's future was indeed bleak. She had no formal education and had only been earning about eighty cents an hour. She rented a small, one-room house badly damaged by the explosion and apparently had only a daughter to help care for her. As an African American, she was also disadvantaged by the racial prejudices of the time.[55]

The second group of surviving employees included men and women who had suffered relatively light injuries in the catastrophe and found alternative jobs elsewhere, often well beyond the waterfront area. More than a few of those individuals were relatively uneducated black workers who performed

the most menial tasks at the chemical plant. Among them was John Lewis, a general laborer with only a third-grade education. Lewis went to work on a rice farm near Bay City, Texas, where he plowed fields, built levees, and planted cereal grain. Elijah Earls, who had a seventh-grade education, also transitioned into agriculture, working as a cotton picker on a farm near Hungerford, Texas. Other survivors in this category had the education and skills to move into more lucrative fields of employment. For example, James Davis, an electrician at the plant, went to work for an electrical company in Bryan, Texas. By 1950, Davis was working for a different electrical company in Waco, Texas.[56]

Other employees who never returned to Monsanto fell somewhere in between these two groups. Gussie Weydert, a white waitress in the plant's cafeteria, was one of them. Weydert had worked at the plant for only three months when the disaster upended her life. At the moment of impact, the waitress suffered a fractured skull and lacerations to her back, left hand, and left eye, leaving her disabled for several months. During that time, she moved to the Chicago area with her husband, where medical specialists helped her recover from the injuries. But with lingering pain in her hand and back, and only an eighth-grade education, Weydert did not take another job until 1951, when she began waitressing in a small town in Missouri. A year later, Weydert was in southern Illinois working at a bottling plant at night inspecting bottle labels. As Weydert's experiences indicate, life remained challenging for many surviving employees long after the waterfront tragedy.[57]

The Texas City disaster offers a good example of what social scientists call a *focusing event*. It was unusually shocking and harmful, and completely unexpected. The news media pounced on it. With stunning photographs and stories of personal tragedies, the media brought the disaster before a national audience. Coming on the heels of World War II, the most destructive conflagration in history, the disaster reminded Americans that the world had changed. The utter devastation in Texas City reinforced the fear that new inventions had slipped beyond human control. Before the war, the American public knew nothing about fertilizer-grade ammonium nitrate. Even the midwestern factories that produced the product had no idea that its chemical compounds could prove so deadly. Clearly, the risks tied to the production and handling of FGAN were too big to ignore. Something had to be done. Policies had to change.

In fact, the disaster ushered in several organizational and regulatory changes designed to prevent, or at least mitigate, similar events in the future. At the local level, it led to the creation of a "green belt" between the industrial and residential areas that gave a new sense of security to neighborhoods near the waterfront. The scope of the tragedy also convinced owners of refineries and other large businesses in the area to establish the Texas City Industrial Mutual Aid Society, to which they contributed funds for the purpose of dealing with any future industrial accidents. Furthermore, it prompted local businesses to set up their own fire and safety departments, and to develop new methods of cooperating with fire departments in nearby towns and cities. These departments learned the importance of using only water to extinguish chemical fires and of spraying water in ways that prevent the scattering of hazardous chemicals. They also learned to equip firefighters with special masks to minimize the dangers of inhaling toxic chemicals.[58]

Public officials across the country took other steps to minimize the dangers of chemicals like FGAN. The federal government immediately set new restrictions on the way FGAN was bagged and stored. New policies required the material to be kept in special containers, well away from other highly combustible chemicals. They also restricted the shipment of FGAN over long distances, required shippers and railroads to clean spaces before loading the material into them, and called for the immediate disposal of any of the material that spilled. Congress gave the US Coast Guard more control over the shipping of any hazardous material into heavily populated areas and specifically prohibited "explosive carriers" from docking in those areas. Cities and states took their own steps to minimize risks. Beginning with New York City, metropolitan areas specifically prohibited ships loaded with FGAN from entering their harbors.[59]

The catastrophe also had major implications for the field of disaster relief. It almost singlehandedly gave rise to the American Association of Blood Banks, for example, an organization that improved methods of collecting and distributing blood, and offered valuable technical and consultation services to emergency response groups.[60] It also motivated emergency responders to develop more effective methods of coordinating relief activities, and identifying and removing dead bodies from disaster sites. The Army Corps of Engineers' approach to dealing with the calamity received high praise from public leaders and became an integral part of the curriculum at the army's

engineering school in Fort Belvoir, Virginia. Perhaps more importantly, the disaster encouraged the federalization of emergency management. Before 1947, large-scale relief and recovery efforts hinged primarily on the ability of local governments to solve local problems. The efforts thus depended on the availability of business loans, insurance policies, and the generosity of churches and charity groups. Congress occasionally passed laws to provide federal funds to help with such efforts, but the process had proven to be cumbersome, uneven, and slow to unfold.[61]

The process changed after the Texas City calamity. Congress started investigating new ways of putting federal resources into the hands of local governments quickly, without delay. As often happens in post-disaster policy making, another major tragedy soon highlighted the need for policy reforms. In June 1947, while Congress debated its options, a sequence of storms caused severe flooding along the Missouri and Upper Mississippi Rivers, laying waste to large parts of the nation's grain-growing belt. The flooding spelled ruin for thousands of families in the area. Democratic Senator James Murray of Montana explained the problem to Congress from a local perspective: "This is not a flood whose debris can be mopped up as the tired housewife sweeps out the mud and filth left in her parlor when the waters receded." It was instead "one of the major flood disasters of all time."[62] Coming on the heels of the Texas City tragedy, this new calamity convinced more Americans that Congress should drop its piecemeal approach to emergency relief and take greater responsibility for disaster victims. The world had simply grown too dangerous.

In the wake of these back-to-back disasters, on July 7, 1947, Republican Senator George Aiken of Vermont introduced a bill that had been in the works for weeks. In short, the bill allowed the president to declare major disasters and send surplus military property to disaster-stricken areas *without* congressional approval. When Aiken introduced the bill, he referred specifically to the Texas City disaster as well as the flooding in the Missouri and Mississippi valleys. Both tragedies, the senator said, represented "good examples" of the kind of calamities that justified immediate federal assistance. Aiken's bill originated in the Senate's Committee on Expenditures in the Executive Department, and the committee's thirteen members supported it unanimously. When Aiken introduced the bill on the Senate floor, no one spoke out against it. On the contrary, Republicans and Democrats alike supported it "heartily," as some of them said at the time. The Surplus Property

Act passed through both houses of Congress quickly, and President Truman signed it into law on July 25, 1947. The president interpreted the legislation retroactively and used it to aid Texas City as well as flood-stricken communities from Minnesota to Missouri.[63]

Over the next three years, as surplus military supplies dried up, the federal government established a more permanent policy for disaster relief. In 1948, Congress first appropriated funds for *general* disaster relief. In doing so, lawmakers again entrusted the president to decide what constituted a disaster and how relief funds should be distributed. Congress further formalized these policies with the 1950 Disaster Relief Act, legislation that received broad, bipartisan support. The act set up an emergency fund of five million dollars (about fifty-five million dollars in 2020), which was ten times the amount of money appropriated for disaster relief in 1948. When introducing the legislation in the House, Congressman William Whittington of Mississippi made clear that Congress intended to augment federal relief funds on an annual basis. Whittington assured states' rights conservatives in the House that relief funds were designed only to supplement and not replace whatever aid state and local governments could render. In defending the legislation, Whittington reminded lawmakers that recent disasters like the Texas City tragedy had overwhelmed the resources of local governments. The magnitude of such disasters could wipe out the resources of any city or state, the congressman warned. Other calamities were coming, he added, and the nation had to be better prepared for them: "No one knows when a disaster like the destruction at Texas City, Texas, will occur. No one knows where a tornado may strike. No one knows when a flood may come."[64]

Court cases arising from Texas City also accounted for the institutionalization of emergency relief policies. More than four hundred Texas City residents and several local businesses sued the federal government after the 1947 tragedy, claiming public officials had acted negligently in directing the nation's postwar foreign aid program. In 1950 the federal district court agreed, and its ruling laid the foundations for a massive restitution program. But to the bitter disappointment of the plaintiffs, the Fifth Circuit Court of Appeals later overturned the ruling, claiming that the federal government had the right to exercise its own discretion in handling matters of national concern. The US Supreme Court narrowly upheld that decision in 1953.

The high court's reversal reinforced perceptions of injustice and thus put

pressure on lawmakers to take still greater responsibility for disaster relief. In 1955, following its own investigation of the Texas City tragedy, Congress passed a bill that provided nearly $17 million ($170 million in 2020) to compensate the disaster's victims for personal injuries, deaths, and property losses. Unlike the Supreme Court, Congress concluded that the federal government was "wholly responsible" for the tragedy and was therefore obligated to compensate victims for their losses. Briefly, the bill authorized the secretary of the army to investigate victims' claims and required the secretary of the treasury to pay all claims that were approved. It also set strict limits on awards according to the extent of losses that claimants sustained. President Dwight Eisenhower signed the bill into law, and victims began filing their claims according to widely advertised instructions. Government attorneys ultimately investigated more than seventeen hundred claims, most of which were ultimately approved. More than half of the beneficiaries were widows, dependent children, and parents of the men and women killed in the disaster.[65]

Post-disaster reforms cannot be measured by legislation alone. In the case of Texas City, the disaster had major implications for the medical profession. Among other things, it inspired improvements in the way hospitals mobilized personnel for the purposes of administering anesthesia and led to significant advances in the field of plastic surgery.[66] Specialists at Galveston's John Sealy Hospital, where more than 350 victims were treated for injuries, performed over a hundred surgeries in the hours and days that followed the disaster. Many of those surgeries involved removing glass and metal particles embedded in survivors' faces, along with dirt, sand, molasses, grease, oil, and even oyster shells. "These wounds were contaminated with every imaginable type of debris," as one doctor put it. Victims at the hospital had suffered an unusually wide range of injuries, which added to the challenges there. "The majority of patients showed, in addition to lacerations, one or more fractures and/or eye wounds, perforations of ear drums, concussion and internal or moderately severe internal blast injury," the same doctor explained.[67]

To deal with the crisis, surgeons at the overcrowded hospital worked in teams, operating on wounds in multiple stages over extended periods of time. Some of their patients had to undergo four or five separate operations, each of which required numerous scar incisions and skin grafts. Seventeen of the patients died as a result of their injuries, nine of whom had gone

through at least one surgery. Looking back on the disaster, the hospital's doctors agreed that they learned valuable lessons about shock therapy, patient consultation, and plastic surgery. Their hard work helped disaster victims, as one surgeon said, "return to a normal life with a minimal amount of disfigurement or deformity."[68] That, too, was part of the proverbial silver lining in this tragedy's dark clouds.

What happened at the Monsanto plant in 1947 was a disaster in the truest sense of the word. As such, it speaks to major themes of this book. It helps us to see, for example, the variety and persistence of workplace disasters in American history. Monsanto's employees were not working-class groups in old industrial centers of the East or Midwest, but rather integral parts of a diverse, modern workforce in an area of the country previously unassociated with large-scale industrial accidents. Though Monsanto slotted its employees into jobs according to their race and gender, the disaster itself made no distinctions in the workplace. It killed and injured highly trained chemists and engineers along with low-skilled dishwashers and day laborers. It also felled scores of people outside the plant, including many in the working-class neighborhood that bordered the waterfront. Such was the nature of disasters in modern work environments: they affected all sorts of workers and often nonworkers, too.[69]

This disaster also reveals the double-sided nature of modern industrial and technical advances. Commodities like nitrate-based fertilizers and synthetic rubber have proven to be exceedingly practical and even vital to national interests, but producing and distributing those commodities created extraordinary risks. Who could have imagined such a powerful industrial accident before 1947? The unintended consequences of material growth had never been so clear. On the road leading into the waterfront area, among the gutted cars and structural debris, someone could have posted a sign that read, "Welcome to the Postwar World!" The sign's small print might have warned, "Caution: Now more than ever, mere mishaps and misunderstandings can obliterate work environments—Entire neighborhoods are now vulnerable to unknown, invisible qualities of new-fangled products!" To be sure, there was nothing inherently sinister about this disaster. No one purposefully destroyed the Monsanto plant or triggered its destruction out of foolishness or greed. When the nitrate-laden steamship beside the plant started smoking, local citizens gathered around the docks to watch the spec-

tacle. Firemen and stevedores methodically doused the vessel with water. Meanwhile, Monsanto's employees kept working. All of these people ought to have been running for their lives. The world as they knew it, with its many modern marvels, was literally about to explode.

3

Airliners Collide over the Grand Canyon!

They may travel at a speed incredible to the savage; but in doing so resign life and limb to the care of others. A broken rail, a drunken engineer, a careless switchman, may hurl them to eternity.

Henry George, *Social Problems*

On the morning of June 30, 1956, Captain Jack Gandy and his five-person crew entered their workplace—a large airliner on the tarmac of the newly renovated Los Angeles International Airport. As Gandy and his crew readied the aircraft for a regularly scheduled flight to Kansas City, Trans World Airlines (TWA) Flight 2, sixty-four passengers subsequently boarded the plane and buckled up for safety, including nearly two dozen TWA employees traveling for free or at discounted prices to join family and friends for Independence Day celebrations. A few minutes later, at the same concourse, Captain Robert Shirley and a crew of five boarded United Airlines Flight 718 and prepared to journey to Chicago. Fifty-three passengers followed them, including a few off-duty UA employees. It was a seemingly perfect day. The two propeller-driven airliners lifted off shortly after ten o'clock, and the 128 people within them settled in for what they assumed would be a long and safe flight.

They were sadly mistaken. At approximately 11:30, the two commercial airliners collided twenty-one thousand feet over Arizona's Grand Canyon, the most stunning landmark of the American Southwest. Apparently, the left wing of United Flight 718 smashed the tail assembly of the TWA plane before its propeller blades ripped across the right side of the plane's fuselage. Without tail stabilizers, TWA's plane turned upside down before plung-

ing into the canyon. The United flight, which lost most of its left wing in the collision, made a sweeping right turn as it careened downward. As it did, the plane's copilot, Robert Harms, sent a final, garbled radio message, later transcribed as: "Salt Lake . . . ah . . . 718 . . . We are going in." There was no sense of hope in the copilot's voice, and understandably so. The planes were flying at nearly four hundred miles per hour when they collided, with only the canyon's abyss below them. When they hit the canyon's steep rock walls, both planes disintegrated on impact, leaving no survivors.[1]

At the time, the midair collision was the worst commercial aviation disaster in history, but it was also one of the most mysterious. The downed airliners—a Lockheed Super Constellation and a Douglas DC-7—were among the world's most advanced commercial aircraft. TWA had introduced its "Super Connie" with great fanfare in 1952. Powered by four 18-cylinder engines, the plane could fly with the prevailing winds all the way from Los Angeles to the East Coast without refueling, and it offered its passengers unmatched amenities, including sleeping berths for those flying first class. The DC-7 was United's answer to the Connie. Introduced in 1953, it could fly nonstop even *against* prevailing winds. Its four engines were as powerful as those on the Connie but even more fuel efficient, and its first-class comforts were also top-notch.[2]

The pilots and crews of the airliners were well trained and experienced. Captain Gandy began working for TWA in 1939. He left the company during World War II to serve in the Army Air Corps but returned when the war ended. By 1956, Gandy had racked up fifteen thousand hours of flying time at the company and had piloted the flight from Los Angeles to Kansas City 177 times. Captain Shirley's work experience at United was equally impressive. Shirley joined United in 1937 and had accumulated more than sixteen thousand flying hours by 1956. For the year leading up to the accident, he piloted planes along the Grand Canyon route. The copilots and engineers of the doomed flights were experienced, too, as were the flight attendants. Arizona's weather was relatively good on the day of the crash, and the airliners were over a sparsely populated area. How, then, could the planes have collided with one another?[3]

Scholars and popular writers alike have taken an interest in the Grand Canyon disaster, and with good reason. To begin with, the tragedy was incredibly shocking and disturbing. It symbolized everything that was frightening about flying. Imagine soaring to twenty-one thousand feet before crash-

ing into a giant moving object and spiraling into the Grand Canyon. What could be more terrifying? The midair collision suggested that airline companies had left their employees and customers at the mercy of malevolent forces, and that government officials had failed to protect the public interest. The air disaster had major, long-term consequences. In its aftermath, Congress created the Federal Aviation Agency (FAA), which oversaw the transformation of the nation's air management system. This is not to say that the event singlehandedly changed the course of aviation history, but rather that it was the prime mover in an air safety reform movement. In effect, it only confirmed what other air accidents had already made clear: that recent industrial and technical advances had left the nation's air control policies badly outdated.[4]

The 1956 air accident sheds light on the paradoxical features of modernity as well as the interrelatedness of disasters and safety reforms. The rise of the commercial airline industry represents one of the great achievements in the history of business enterprise. The industry's origins are lodged in a set of remarkable technical advances and scientific breakthroughs as well as public policies that promoted economic development. But those innovations and policies raised the possibility of great tragedies. In other words, they came with risks. Yet such tragedies often create new opportunities to promote public safety. In the field of aviation no less than other areas of economic life, real reform required support from disparate groups with disparate interests and thus demanded cooperation and compromise. Transforming the nation's air control system was a monumental challenge. It demanded political leadership, technical expertise, and a real commitment to institutional and industrial change.[5]

This tragedy also draws attention to the rapidity of change in the American West after World War II. During the war years, Southern California became a nerve center of industrial activity, especially in the field of aviation. A host of large corporations established factories and headquarters in the greater Los Angeles area, including the nation's largest aircraft manufacturers. These and related developments fueled business activities across the economic spectrum and lured tens of thousands of people to the area. This growth warranted a major expansion of the Los Angeles airport and the subsequent scheduling of additional commercial flights in and out of the facility. The Grand Canyon disaster reflected the shadowy side of these changes.[6]

↓

Midair collisions have never accounted for more than a tiny percentage of air disasters, but they have a long and multifaceted history. The first fatal collisions between aircraft in flight occurred in the early twentieth century, when pilots flew planes with two stacked wings. During World War I, several of these biplanes collided over western Europe in the heat of aviation duels known as "dogfights." The first known collision involving a biplane and an American commercial aircraft happened on April 21, 1929, near San Diego, California, when a young US Army pilot hit a single-winged craft carrying three passengers. According to witnesses on the ground, the army pilot had been "stunting" near the larger passenger plane just before his wings clipped its cockpit. The larger plane then pitched upward, rolled over, and crashed to ground. No one survived the disaster, including the army pilot.[7]

In the decade after the war, private planes were involved in more than fifty midair collisions.[8] The worst of these accidents occurred on November 1, 1949, when an Eastern Airline's flight approaching Washington National Airport slammed into a military fighter plane three hundred feet above the Potomac River, killing fifty-five people.[9] The *Washington Post* described the crash site in gruesome detail: "The impact of the crash was so great that bodies were hurled more than 150 feet." "Many of the bodies were missing arms and legs, and at least two were headless."[10] Another ghastly disaster unfolded on January 12, 1955, when a TWA airliner hit a small private plane shortly after leaving Cincinnati, Ohio, killing fifteen people, everyone on both planes.[11]

The number of times pilots narrowly avoided colliding in midair was more alarming than the disasters themselves. According to pilot reports, in the mid-1950s there were nearly four "near misses" a day. In a quarter of these incidents, pilots passed within a hundred feet of one another, or close enough to read the identification numbers on a plane's fuselage. Many of these misses involved military aircraft and commercial planes, and required pilots to employ hair-raising maneuvers to avoid disaster.[12] Captain Robert Stone of United described the experience as "spine-chilling." "Surviving a near-miss is one of the most personal things that can ever happen to a pilot," he explained.[13]

Two near disasters in 1956 foreshadowed the Grand Canyon tragedy. On March 30, the pilot of a TWA plane flying eastward over Pittsburgh unexpectedly encountered a Canadian airliner heading southward at his own assigned altitude of sixteen thousand feet. To avoid a collision, the pilot jammed

the nose of his plane downward and flew just below the Canadian craft, or close enough to read the small lettering on its fuselage. According to reports of the incident, the Canadian pilot was flying directly into the sun and never even saw the TWA plane.[14] A week later, the pilots of two airliners en route to Washington, DC, found themselves in dangerous proximity as they approached the nation's capital. They realized the closeness of their positions after one of the pilots radioed air traffic controllers on the ground and said he felt the "prop-wash" of another plane. One plane was directly above the other when the controllers finally managed to separate them.[15]

The recent growth of the commercial airline industry largely explained these problems. Before World War II, aircraft production was still a relatively small industry that generally relied on skilled workers familiar with concepts of flying and navigation. The Army Air Corps, the predecessor of the US Air Force, had only a thousand planes, many of which were not suitable for aerial combat. The war changed everything. When Germany invaded France in 1940, President Franklin D. Roosevelt called for the construction of fifty thousand military planes within the year. Federal agencies financed the expansion of aircraft production facilities across the country, a move that helped to turn aircraft manufacturing into a mass production industry where large numbers of relatively unskilled workers performed simple, repetitive tasks. Large corporations such as the Boeing Company, Douglas Aircraft, and the Lockheed Company emerged from the war with gigantic factories and workforces as well as top-notch research and development teams and plenty of capital. The companies quickly converted military transport planes into commercial airliners but soon designed new planes more suitable for peacetime purposes.[16]

By the mid-1950s, there were more than a hundred thousand airplanes in the nation, roughly two-thirds of which were civil aircraft. The nation's four leading airline companies—American, TWA, United, and Western Airlines—owned about fourteen thousand planes that ferried passengers and cargo to large metropolitan areas across the country, and more than a dozen smaller firms had fleets that carried passengers and cargo to less populated communities. Private citizens had planes, too, and had often been involved in midair collisions. American airports were the busiest in the world. Chicago's Midway Airport alone handled more air traffic than airports in London, Paris, and Rome combined. Even smaller airports in midsized cities such as Chattanooga, Tennessee, were as busy as most European facilities.

In a word, the skies were crowded. On clear days, people spotted as many as two hundred commercial planes passing over New York City.[17]

Developments in Los Angeles mirrored the larger patterns. During the war years, aircraft manufacturers established several large production facilities in the greater Los Angeles area. The Lockheed Company, for example, built facilities in Burbank, Long Beach, and Santa Monica. Four other producers of planes—Boeing, Douglas, Hughes Aircraft, and Northrop—had plants nearby. Each of these companies relied on parts and services provided by scores of smaller companies in the metropolitan area. At the fastest pace possible, using the most advanced techniques of mass production, these various business enterprises fabricated fuselages, wings, cockpits, and other necessary parts of military planes. Collectively, these companies produced roughly a quarter of the nation's military planes, including heavy bombers, transport planes, and smaller fighter craft. These companies employed thousands of engineers and scientists who helped turn the Los Angeles area into the center of the nation's still emerging aerospace industry.[18]

By the late 1940s the area's rising population justified a major expansion of the Los Angeles Airport, upending the look and feel of entire neighborhoods along the western edge of the greater metropolitan community. When the war began, the city's airport was little more than a bean field with a few dirt landing strips. It did not extend west beyond Sepulveda Boulevard, a major north-south route that formed part of California's Pacific Coast Highway (State Route 1). City leaders mounted a successful campaign to modernize the airport in 1949 that began with the rerouting of Sepulveda and the construction of additional runways, airplane hangars, and a control tower on a large parcel of land that stretched west of the old airport, where the traffic had once flowed. Meanwhile, construction began on a massive underground project that re-created the old boulevard beneath the airport's new runways and taxiways. For four years, builders blasted and pounded the ground, destroying or disrupting entire neighborhoods. In 1953, when the new tunnel opened, Los Angeles boasted of having one of the nation's most modern airports. Major airline companies soon turned the airport into one of nation's busiest air facilities. On the eve of the Grand Canyon tragedy, United offered more than fifty regular weekday departures from the airport, and three of its main competitors—American, TWA, and Western—offered about half that many. Airliners of smaller companies like Bonanza Air Lines and Pacific Southwest also flew in and out of the airport on a daily basis.[19]

The expansion of the Los Angeles Airport closely paralleled major technological advances in aviation. Boeing and Douglas Aircraft had recently introduced planes with advanced air-cooled engines, air pressurization devices, and noise-canceling plastic insulation as well as better cockpit instruments such as altimeters, compasses, and airspeed indicators. Their innovations allowed planes to fly faster, farther, and higher, and with more passengers on board. They also made air travel more comfortable. By the mid-1950s, airliners could fly more than three hundred miles per hour at altitudes topping twenty thousand feet. Some could seat more than a hundred passengers and had a range of more than forty-five hundred miles. Businessmen and vacationers therefore chose air travel over alternative modes of transportation. Between 1945 and 1955, when the nation's population rose from roughly 140 million to 165 million, the annual number of passengers on domestic airliners jumped fivefold, to more than 38 million.[20]

Unfortunately, the nation's air management system had not kept up with these developments. Congress created the basic structure of that system with the 1938 Civil Aeronautics Act, which transferred government responsibilities for civil aviation to the Civil Aeronautics Administration (CAA). The CAA was originally an independent agency charged with developing new air routes, operating air traffic control facilities, and regulating business practices in aviation. By the 1940s, however, President Roosevelt's administration concluded that those responsibilities were too large for a single agency and set up the Civil Aeronautics Board (CAB) to regulate aviation practices and investigate the causes of air accidents. The administration placed both agencies under the control of the Department of Commerce.[21]

In 1946, a long-standing government-industry advisory group, the Radio Technical Commission Aeronautics (RTCA), submitted a report to Congress calling the nation's air management system cumbersome and inefficient. As the report noted, military planes often encroached upon civilian airspace without informing the CAA, and the agency was woefully underfunded and understaffed. "The tools available to the CAA to discharge [its] responsibility are marginal even by pre-war standards," the report stated. The agency's controllers were so overworked that they had difficulty staying abreast of aircraft positions. "Traffic controllers are struggling valiantly to handle an increasingly difficult situation [but] at present, the only position information available to controllers . . . may be in error by many miles."[22] The report

Air traffic controllers tracking planes along "airways," circa 1950. To reach their final destinations, airline pilots often had to veer off these paths and fly planes into uncontrolled airspace, where they flew according to visual flight rules. Getty Images

concluded that the nation desperately needed to improve its air traffic control methods.

Methods of flight control were surprisingly simple in these years. At mid-century the CAA employed about fifteen hundred controllers. These controllers worked at 120 airport towers and 25 en route flight stations. Their jobs were not especially complicated, but mistakes could cost lives. Controllers in airport towers reviewed commercial flight plans and assigned planes to positions along established "airways," which were essentially blocks of airspace about ten miles wide and fifty miles long. In doing so, they tried to place a minimum of ten minutes of flying time between planes traveling in

the same direction and at least a thousand feet of vertical separation between those heading in opposite directions. Using assigned radio frequencies, the controllers then cleared planes for takeoff and tracked them as they
took to the sky.[23]

The other group of controllers, those at en route facilities, monitored
flights as planes passed through their jurisdictions. To do so, they sat side
by side at desks facing a rack of slanted metal trays, each of which represented a thousand feet of altitude within a section of an airway. Working
with designated radio operators at aeronautical ground stations, the controllers received regular flight position reports from pilots as planes passed
over designated radio checkpoints. The controllers wrote their own reports
on strips of papers and placed them on the appropriate trays. So long as
these "fixed postings" were on different trays, controllers assumed flights
were sufficiently separated. When flights passed beyond the area an individual controller monitored, he handed his posting to the coworker seated next
to him, who continued to track the flight until it moved beyond the facility's
jurisdiction. Upon notification, controllers at other facilities monitored the
flight until airport tower operators cleared it for landing.[24]

This air management system was problematic on several levels. To start
with, controllers only monitored *established* airways, which accounted for
less than a quarter of the nation's airspace. For planes to reach their final
destinations, their pilots often had to enter "uncontrolled" airspace where
they followed Visual Flight Rules (VFRs), otherwise known as the "see-and-
be-seen" principle. In this mode of flying, pilots had to watch for approaching planes and take evasive action to avoid collisions. These rules seemed
reasonable in the days when "barnstormers" in open-air cockpits ruled the
skies, but they were ill suited for the age of commercial aircraft. The speed
of modern airliners made it difficult for pilots to identify approaching planes
in a timely fashion, especially if those planes were coming out of clouds or
in front of the blinding sun. As time and motion studies had recently shown,
it took pilots at least six or seven seconds to change the direction of an airliner after spotting a collision object. But it took even less time for planes on
the horizon to overtake the airliner. The recent development of new military jets only exacerbated the problem. Traveling at more than five hundred
miles per hour, these new planes, powered by "turbojets," could change from
a speck on the windshield to an impending disaster in about five seconds.[25]

Flying "off airways" was dangerous for still another reason: manufactur-

ers had not designed the pilot's workspace with a safety-first principle in mind. Surprisingly, the cockpits of commercial airliners offered only a partial view of the surrounding skies. Some cockpits provided pilots with a mere 31 percent angle of vision looking forward and an even narrower side view. And unlike their predecessors in open-air cockpits, airline pilots could see nothing directly above, below, or behind them. To make matters worse, they had to stay above cloud tops and dodge thunderstorms, all while being able to communicate with their crews and sometimes their passengers. Meanwhile, their flight engineers had to engage in a variety of work routines that compromised their ability to detect approaching aircraft. They needed to monitor instrument panels, for example, a task that involved selecting fuel tanks, resetting altimeters, and changing radio channels.[26]

Airline pilots had long complained about these conditions. Acting largely through their union—the Air Line Pilots Association (ALPA)—the group waged a sustained postwar campaign to raise air safety standards across the nation. In the early 1950s, when ALPA had approximately nine thousand members and forty contracts with airline companies, the union's chief representatives informed Congress that rapid business and technological developments had made it almost impossible for pilots to ensure the safety of their crews and passengers. ALPA's first president, David Behncke, appeared before congressional committees several times in these years and forcefully chided lawmakers for failing to fund air safety programs. Behncke accused federal officials of being so beholden to industrialists that they had ignored serious safety problems, suggesting that CAA policies reflected political expediency more than common logic. The labor leader therefore urged Congress to establish a new independent agency to promote air safety.[27]

The CAA defended itself against these charges to the best of its ability. In 1955, CAA chief F. B. Lee wrote a candid letter to ALPA members acknowledging big problems in the realm of air management. As Lee told the members, the number of flight position reports posted by controllers at CAA facilities had more than doubled since 1945, jumping to sixteen million. In that same ten-year period, he noted, the federal government had given the military additional airspace to test weapons and planes while also prohibiting commercial flights over certain urban areas, making it even harder for the CAA to protect commercial airliners. "There are now some 450 restricted areas scattered across the country," Lee complained. "This number is just about double the number back in 1948." The construction of tall radio and

television towers near airports created still more challenges for air traffic controllers, as did the development of faster-flying military planes.[28]

The CAA chief also reminded the union's members that his agency had recently taken several steps to address air safety concerns. It had explored new ways of presenting flight information at en route control facilities, for example. As Lee explained, the CAA wanted to replace the standard flight data strip with new panoramic-type displays that would give controllers a more complete perspective of air traffic patterns. More importantly, the agency had introduced new military-type radar surveillance technology at three airports located near military installations and planned to introduce additional radar units at a dozen other airports within the next few years. The CAA also planned to equip en route control facilities near large urban areas with this new technology.[29]

Secretly perfected during the war, radar technology harnessed electromagnetic energy to detect the presence of objects in the air, with major implications for commercial aviation. Simply put, radar devices transmitted pulses of electromagnetic waves in a straight line. If the waves met any objects, such as a thunderstorm or the fuselage of an airplane, they bounced off the objects and returned in weakened state to the place where they originated. The speed at which the waves returned helped to determine the location of the objects they encountered. During the late stages of the war, the armed forces deployed many rotating, ground-based radar devices to help spot enemy aircraft, and in the early postwar years they introduced "radar scopes" in the cockpits of military planes to help pilots detect approaching aircraft and find smooth corridors through rainstorms.

By the mid-1950s, airline companies had also begun investing in radar technology. In late 1954, United's president, W. A. Patterson, described airborne radar as "one of the great technological advances of the last 10 years." His own company, Patterson said, planned to install radar devices in its own planes within the near future.[30] The pilot's union seized on those words and urged other companies to follow United's lead. "We hope that other carriers will shortly follow United Air Lines and make airborne radar available to their pilots," the union's newly elected president, C. N. Sayen, said at the time. "The inclusion of airborne radar in the cockpit will help the pilot, by visual reference to a radar scope, to determine the location of thunderstorm cells, heavy precipitation, turbulence and icing areas and will help to insure

proper terrain clearance."[31] As Sayen no doubt realized, the airline industry was entering a new age in the history of aviation.

The rapidly changing world of air travel had implications for the US Bureau of the Budget, the agency responsible for reviewing funding requesting and advising the president on matters of spending. Throughout these years the bureau struggled to maintain fiscal restraint while addressing growing concerns about air safety. In 1955, facing mounting pressure to increase spending on air control facilities, the bureau formed an advisory group to study the aviation industry's long-range needs.[32] Headed by William B. Harding, a former air force officer and an investment banker, the group interviewed representatives from airline companies, government regulatory agencies, and the armed forces before submitting its report. To put it mildly, Harding's report blistered the federal government for not doing more to ensure the safety of air travelers. As the report concluded, the nation's airspace was "overcrowded." The pace of technological change in the field of aviation had been fast and furious, and the nation's air facilities had not kept up. "In many important areas, the development of airports, navigational aids, and especially our air traffic control system is lagging far behind both aeronautical development and the needs of our mobile population," the report found.[33] Air congestion had limited military training operations at a time when military readiness was more important than ever. Presciently, Harding envisioned a deadly midair collision involving two commercial airliners: "The risks of mid-air collisions have already reached critical proportions. Fortunately, so far, none have been between two of our largest transports when fully loaded."[34]

In response to Harding's report, President Dwight Eisenhower asked former Major General Edward P. Curtis to recommend ways of modernizing the flight control system. As the president realized, Curtis had a deep understanding of the aviation sector as well as a sterling reputation within the military. During World War II, he helped direct strategic air operations in Europe and Africa that hinged on the successful and rapid integration of new navigational aids into the armed forces. The general assumed his new duties in March 1956, with a three-man staff and a budget of $450,000 (about $4.3 million in 2020). He was in the first phase of the project when the two airliners collided over the Grand Canyon.[35]

↙

The Grand Canyon disaster confirmed what critics of the air management system had been saying for a decade: the system was antiquated. As post-crash investigations revealed, pilots Gandy and Shirley had followed all the rules of aviation. Before their planes left Los Angeles, the pilots filed detailed flight plans with air controllers. Gandy proposed flying the TWA craft along a well-traveled route to Kansas City at a cruising altitude of nineteen thousand feet, and he designated specific towns and landmarks along the way as his radio checkpoints. Shirley proposed a different route to Chicago, at twenty-one thousand feet, and marked his own radio checkpoints. Controllers approved both flight plans, and the two pilots followed normal departure procedures. Gandy's flight lifted off the ground at 10:03 a.m., three minutes before Shirley's flight, and the two planes reached their respective cruising altitudes without problems.[36]

As was often the case, controllers agreed to modify flight plans once the planes were airborne. About twenty minutes into his flight, Gandy radioed TWA ground operators and asked to take his plane from nineteen thousand to twenty-one thousand feet owing to wind and cloud patterns. Ground operators relayed Gandy's request to controllers in Los Angeles, who in turn asked for advice from their counterparts in Salt Lake City: "TWA 2 is requesting two one thousand, how does it look?" Controllers in Salt Lake City had jurisdiction over the airspace into which both flights were heading and saw that United Flight 718 was already at twenty-one thousand feet and on a crossing path. They thus advised against Gandy's proposal. "Their courses cross and they are right together," Salt Lake pointed out. Controllers in Los Angeles therefore denied Gandy's request, but the pilot then asked to climb at least a thousand feet above the cloud tops, where visibility was better. Los Angeles approved the request but warned Gandy that United's flight was at twenty-one thousand feet and on a converging path: "Advise TWA 2 his traffic is United 718."[37]

A few minutes later, when Gandy reported his position over Daggett, California, he was twenty thousand feet in the air. After passing Daggett, however, in accordance with his original flight plan, Gandy steered his plane leftward into uncontrolled airspace, where he was free to modify his altitude without notifying air traffic controllers. The pilot made his last radio report at 10:55, over Lake Mohave on the Nevada-Arizona border, noting that he had leveled off at twenty-one thousand feet, an altitude he had previously requested and controllers had subsequently not allowed. At that moment,

Shirley's plane was about a hundred miles to Gandy's right. Shirley reported his position at 10:58 over the border town of Needles, California, after which he too turned his plane leftward and entered uncontrolled airspace.[38] Both pilots, then, were heading toward the Grand Canyon on converging paths at the same altitude, within airspace where there were no binding rules of aviation. In other words, they alone had the responsibility of avoiding other planes.

Whether either pilot ever spotted the other is unclear. The two planes collided about eighty-five miles east of where their flight paths should have crossed, suggesting the pilots might have veered off course to dodge clouds or rainstorms. As a pilot for American Airlines later told crash investigators, he had seen scattered clouds a few miles south of the crash sites only an hour after the air disaster. The tops of the clouds rose as high as twenty-five thousand feet, the pilot recalled. In his words, the cloud buildup "looked very much like a thunderstorm."[39] People on the ground corroborated the pilot's account. Some remembered light rain in the area, and even thunder and lightning. The press tended to disregard the weather's impact on the crash, speculating instead that Gandy and Shirley may have adjusted their flight paths to give passengers an unobstructed view of the Grand Canyon. Pilots were known to do just that when flying over scenic areas, and the spectacular canyon was the crown jewel of the western landscape. Whatever the circumstances, neither pilot violated any rules of aviation. Gandy was only five miles off his proposed flight path at the time of the collision, and Shirley was only twenty-five miles off. The distance was reasonable for pilots flying in uncontrolled airspace, especially in inclement weather.

It took hours to verify the sad truth of what had happened. When the planes failed to report their positions to the air traffic control center in Salt Lake City, the supervisor on duty there, Stuart Halsey, realized the gravity of the situation. Halsey followed the protocols. He telephoned radio ground stations and control centers in the region to report the planes missing. He also notified TWA and United, whose dispatch offices in Los Angeles had been monitoring the flights. Meanwhile, using all available radio frequencies, his Salt Lake City control center attempted to contact the missing pilots. The news that radio operators in the Salt Lake area had picked up a garbled but unidentified message took on greater meaning. Was it a distress call, a Mayday message? Shortly before one o'clock in the afternoon, about two hours after last hearing from the missing planes' pilots, the CAA issued a

Artist's conception of the tragic air disaster over the Grand Canyon, June 30, 1956. Two commercial airliners collided at an altitude of twenty-one thousand feet and then spiraled into the canyon. No one survived. Getty Images

missing aircraft alert and initiated an intensive search and rescue mission involving the US Air Force and Forest Service together with the Arizona and New Mexico State Police. About a hundred planes subsequently took to the skies in search of plane wreckage between the Colorado River and Winslow, Arizona, a huge area where controllers assumed the planes must have gone down.[40]

Two brothers—Palen and Alfred Hudgin—were the first people to spot the crash sites. The Hudgins were pilots who owned a small aerial sightseeing company that consisted of a few lightweight single-engine planes, two dirt airstrips, a hangar, and a small office building on the canyon's southern margins. On the day of the crash, at 12:30 p.m., Palen was giving three tourists a bird's-eye view of the canyon when he noticed wisps of black smoke over the canyon's eastern rim, near the confluence of the Colorado and Little Colorado Rivers. The pilot initially assumed the smoke came from a brush fire, perhaps started by a lightning strike the previous day when storm clouds had passed over the area. But after hearing a late-afternoon radio report about the missing airliners, he and his brother flew back to the area where he saw the smoke and looked for signs of the disaster. As darkness fell, the men saw a combination of plane parts and bodies scattered above the west bank of the Colorado River, including a tail assembly with TWA markings. About a mile away, they noticed black smoke rising from a precipitous ledge high above the river. The brothers then flew back to their small airport and reported the discovery to TWA's flight control center in Kansas City.[41]

At daybreak the Hudgins returned to the area and pinpointed the sites for authorities. As later reports indicated, United's plane had smashed violently into the south face of Chuar Butte, a massive rock formation that towered above the western bank of the Colorado River. The plane hit the butte's steep cliffs at an elevation of thirty-four hundred feet, or about halfway up the prominence. A large part of its fuselage lodged in a deep fissure about twenty-six hundred feet above the river, but debris and dead bodies were all over the rocky slope. The TWA flight had slammed into the northeast terrace of Temple Butte, another massive prominence above the Colorado River. Most of the TWA plane ended up on the slopes of the Butte about five hundred feet above the river, along with the bodies of its crew and passengers.[42]

The canyon's rugged terrain and unpredictable air currents made the task of identifying victims and recovering plane parts unimaginably difficult

and dangerous. As TWA's vice president, John Collings, explained, "Had we had set out to deliberately find the most inaccessible place in the United States in which to have an airplane accident, I don't believe we could have found a place more inaccessible than where our plane and United's crashed in the Grand Canyon."[43] Collings initially considered putting boats in the Colorado River and "shooting the rapids" to reach the TWA site, but getting victims and remnants of the planes back on boats seemed impossible. The two-way journey would also have taken more than a week. Authorities ultimately decided to use helicopters to reach the site, but strong air currents prevented them from doing so for two days.

Rescue workers had an even harder time getting to United's crash site. The top half of Chuar Butte was so steep that the world's best mountaineers dared not rappel down it. Getting to the site from the Colorado River looked easier, but climbers abandoned that strategy after two days of fruitlessly hammering spikes into cliffs and taking hair-raising steps skyward. The rescue operation eventually involved landing a helicopter on a small rocky point about ten feet above the site and lowering a team of highly trained specialists from Switzerland to the site from there. Once the Swiss team reached its destination, it dropped ropes to US Army crews positioned at a lower spot on the butte and began sending down whatever it retrieved, including body parts.[44]

The rescue efforts led to the identification of only 30 of the 128 victims, the vast majority of whom had been on the TWA flight. Most of the bodies from United Flight 718 were either too deeply embedded in the canyon's walls to dislodge or they disintegrated, becoming mere ashes when rescuers tried to lift them. Others came to rest in out-of-reach crevices and chimneys along cliffs. In the end, rescue workers retrieved the remains of twenty-nine people on the flight, almost none of whom were actually identified. Authorities later buried the unidentified body parts from United's crash site in a graveyard within the Grand Canyon National Park. Remains from the TWA crash went to a common grave in Flagstaff, Arizona.[45]

A US Army officer charged with processing human remains from the TWA crash later described the macabre experience. As he explained, army helicopters brought pouches of body parts to a dirt airstrip on the canyon's rim, where military personnel transferred the pouches to a plane and flew them to Flagstaff. Upon landing, the team put the pouches in a large hearse and drove them to a makeshift morgue. Authorities then prepared a gravesite at

Rescue workers retrieve tail assembly of TWA Flight 2, Grand Canyon, 1956. Box 145, TWA Records-K0453, State Historical Society of Missouri Research Center, Kansas City

a local cemetery large enough to bury seventy caskets, one to represent each person that had been on Flight 2. "We blasted a hole big enough to hold 70 caskets," the officer in charged explained. "Then we built a platform over the big hole, got artificial grass to cover the platform, got a big circus tent erected to provide shade for the mourners and had a big service at the gravesite." Sadly, many of the caskets that were buried at the site were nearly empty. As the officer later explained, "We had to divide what pieces we'd recovered, so that the press couldn't report we didn't have something in each casket."[46]

⩩

By July 7, when authorities buried the caskets in Flagstaff, newspapers across the country had reported the air disaster's most grisly details. Perhaps more importantly, they also raised pressing questions about the effectiveness of air traffic control methods. A postcrash editorial in the *Boston Globe* reflected the general message: "Why were these two heavily loaded airliners dispatched on an identical flight course only 3 minutes apart? Why were not divergent, or at least parallel courses, 30 or 40 miles apart, assigned them?" "Wasn't this far too hazardous?" The editorial called on Congress to regulate the use of airspace more scrupulously, and immediately. "This catastrophe reveals an urgent need for newer and more rigorous precautions," its author concluded. "The whole air traffic control system needs an overhaul."[47]

In fact, Congress had already taken steps in that direction. After the crash, the legislative body with jurisdiction over the field of aviation—the House Committee on Interstate and Foreign Commerce—formed a twelve-man investigative subcommittee. Chaired by Arkansas Democrat Oren Harris, the subcommittee immediately traveled to Las Vegas and held public hearings at one of the metropolitan area's most iconic landmarks, the Sands Hotel. More than a dozen leading figures in the field of aviation testified there, including representatives from the pilot's union and leading airlines as well as the CAA, CAB, and armed forces. Others spoke to the subcommittee in follow-up sessions in Washington, DC.

Like the disaster itself, these hearings drew widespread attention to the dangers of flying. One of the first people to testify in Las Vegas, CAA Administrator John Garrison, admitted that his agency had difficulty managing air traffic. As Garrison reminded public officials, air safety hinged on pilots detecting and avoiding approaching planes, even when they were in established airways, where Instrument Flight Rules (IFRs) applied. "Other aircraft may be flying indiscriminately throughout that area," Garrison explained.[48] Oscar Bakke of the CAA acknowledged the dangers of flying "off airways," where air travel was largely unregulated. "The total capacity of the air traffic system is adequate to handle only a small percentage, it has been said about 15 percent, of all the traffic that moves in the United States airspace."[49] These and other statements from CAB representatives reflected the limitations of air management strategies and technologies.

These limitations became much clearer after the testimony of TWA pilot Jasper Solomon, who had flown one of his company's Super Constellations

Congressmen and mountain climbers discuss search and rescue operations at the Grand Canyon, July 8, 1956. United Flight 718 slammed into the steep walls directly behind the group. Associated Press

over the Grand Canyon the very morning of the crash. Unlike CAB representatives, Solomon represented the workers in this narrative, and his words thus held special significance. As he told the Harris subcommittee, pilots who worked for different companies did not share a common radio frequency and thus relied on a cumbersome relay system to get information about other flights in their vicinity. "You have to call the airways radio station and ask them for any information that you might need," Solomon explained. Solomon often had trouble contacting radio operators working for other airline companies, as they were not always tuned to his airway's frequency. The pilot suggested that Captain Gandy would not have been able to notify Captain Shirley of his altitude and location.[50]

Solomon's testimony raised yet more questions about the safety of air travel. He admitted, for example, that pilots sometimes changed their flight paths to give passengers a better view of scenic landscapes, especially the Grand Canyon. Truth be told, Solomon had altered his own path that morning to showcase the magnificent site. "I went to the north rim, and made a right-hand turn and flew across the canyon to the south rim." At the time, he remembered, the skies were clear enough for his passengers to see the ground.[51] Perhaps, then, Gandy and Shirley had purposefully flown off course, at least slightly. But Solomon also suggested that the poor design of airline cockpits explained the crash. After all, the design severely limited a pilot's field of vision. "It is possible or conceivable to get into a situation where you would have one plane directly over the other and they could fly for hours and never see each other," he told lawmakers. "At some speeds," Solomon added, "you do not have enough time after seeing an aircraft to make a move. Your visual distance is not great enough."[52]

The pilot's union backed up these statements. Speaking for the union, President Sayen blamed lawmakers for putting pilots in such dangerous positions, arguing that Congress had not adequately funded the CAA and its air control facilities. "We have had a situation where appropriations for air-traffic-control development and the CAA generally have been decreasing since about 1950 while aircraft movements in this country both in numbers and in complexity have increased quite largely," he explained. CAA control towers at major metropolitan airports, he added, were so grossly understaffed that the agency had come to rely on controllers with little work experience. "In some centers where the traffic has developed very rapidly, they have not been able to recruit and train personnel fast enough," Sayen said. The labor leader ended his testimony by demanding immediate and profound changes to the nation's air management system. "We are badly in need of a modernization and improvement of the entire system of air-traffic control," he concluded.[53]

In a moment of labor management cooperation, industry spokesmen agreed with union leaders. The head of the airline industry's leading employer organization, Milton Arnold of the Air Transport Association (ATA), echoed labor's point of view. "During the past several years," Arnold testified, "the expansion of aviation has far outdistanced the CAA's ability, under existing appropriations, to maintain and provide an adequate level of experienced personnel to absorb the increasing duties." Arnold urged Congress

to improve the salaries of air controllers, whom he described as the "back-bone" of the air management system. As the industry spokesman noted, those salaries had not increased in several years, and congressional conservatives were currently trying to roll them back to earlier levels. "Unbelievably as it might sound," Arnold complained, "there is now a plan approved by the Civil Service Commission which would reduce the base salary of controllers." Arnold described the jobs of air controllers as "nerve-wracking." "The pressure and strain of the work itself is, in fact, showing its effect in the physical and mental health of controllers." The shortage of skilled employees in airport control towers was real. In some places the shortage had forced experienced controllers to work eight-hour shifts without lunch or even rest periods.[54]

Arnold pushed other proposals, too. For example, he called for the addition of new high-intensity lights along the runways of small airports to help pilots when landing and the establishment of new airways designated specifically for long, high-altitude flights. According to Arnold, the design of the existing airway system was not well suited for new commercial aircraft: "The airway network is laid out from city to city, logical enough for short-and medium-distance flights, but antiquated as a cow path for Super Constellations and DC-7's flying coast to coast." Echoing the words of pilots and their union, Arnold also urged Congress to find better ways of relaying flight information to commercial flights, calling existing methods "slow and cumbersome." "For communications, we are still relying heavily on voice telephone and voice radio, which is like a party-line system, where one must listen to all calls to be sure he will recognize those intended for him."[55]

The statements by Sayen and Arnold carried considerable weight. Both men were speaking for powerful interest groups in the policy-making process, and thus they were well positioned to shape the direction of policy debates. Both also approached policy making with a sense of authority and dedication, and they demonstrated formidable skills as policy advocates and negotiators. Unlike the millions of air travelers that boarded commercial airliners each year, Sayen and Arnold had the time and resources to promote air safety. They could pinpoint the central concerns of their respective coalitions and had a better understanding of air safety problems than perhaps any congressmen at these hearings. Yet the men were not specialists trained in branches of science and engineering. To get a better sense of the limits and promise of new technologies in aviation, Congress relied on testimony

from other individuals who understood how new navigational devices could solve pressing air traffic problems.

The testimony of C. C. Furnas, an engineer from the Department of Defense, was particularly insightful. Furnas told Congress that the military's newly developed beacon system could track commercial flights more effectively than existing technologies. This system consisted of ground-based "secondary surveillance radar" with multidirectional antennas that scanned the skies emitting signals, to which transponders on planes replied. The replies gave military controllers precise information about the speed and the location of planes and thus helped to keep planes separated in air.[56] Furnas also brought attention to new computers that the military used to help controllers transfer information to one another almost instantaneously, without using telephones or teletype machines. The computers also helped controllers store and retrieve information more quickly and could solve complex mathematical problems in fractions of a second. Computer technology, he added, could eliminate many distracting day-to-day activities in control facilities. Within a broader perspective, Furnas suggested the machines would eventually replace many of the controllers' job responsibilities. "It is possible to replace, at least in part, the mind and memory of the controller with very rapid computers, mathematical computers, and magnetic storage devices," he told lawmakers.[57]

Speaking for the CAA, James Pyle told lawmakers that his agency had already begun integrating new military technology into its air management system. For example, the CAA had recently purchased military-style radio transmitters and receivers capable of carrying radio messages over high-frequency waves that were relatively uncongested and virtually static-free. It also invested in long-range radar that improved air control methods. As Pyle explained, radar technology was truly transformative: "Radar provides a means for the traffic controller to 'see' the aircraft under his control rather than having to visualize their positions by means of written estimates of positions arranged on a display rack." The CAA expected radar not only to make the skies safer but also to allow more planes to fly within a controller's jurisdiction. "Without radar, the traffic controller must keep aircraft separated from each other by comparatively large amounts of airspace," Pyle told lawmakers. "With radar, the traffic capacity can be increased drastically, on the order of 4-to-1."[58]

William Andrews of CAB's Bureau of Safety Investigation offered insights

into other promising new technologies. Most importantly, Andrews referenced the Ryan Flight Recorder, or flight analyzer. Developed by the University of Minnesota and the General Mills Corporation, these "black boxes" recorded the speed, altitude, and compass direction of planes and were made to withstand power failures and even plane crashes, and could therefore help investigators determine the causes of air accidents. As Andrews noted, however, the devices were still too heavy and complicated to use in commercial aviation. Moreover, they would not likely survive disasters like the Grand Canyon tragedy. "The wreckage was so terrific, and the fire was so terrific, that I doubt seriously whether any instrument would have been saved," the CAB spokesmen explained.[59]

The Harris subcommittee continued to explore air traffic control methods until January 1957, when it finally submitted a report to Congress. The report concluded that air traffic control methods were indeed outdated. Planes were taking off and landing every few seconds at airports whose control towers were understaffed and poorly equipped. Problems were mounting every day. The report predicted that by 1965 the nation's annual number of air travelers would jump from 38 to 93 million and the speed of airliners would soar from 300 to 600 miles per hour. The time for new regulations had come. "The United States in the past 30 years has built an airways system unmatched anywhere else in the world," the report concluded. "Still, this vast, highly complex system is not adequate in that it is incapable of meeting the air traffic demands."[60]

The support for greater regulation grew stronger in April 1957, when former Major General Edward Curtis submitted his much anticipated *Aviation Facilities Planning Report* to President Eisenhower. After consulting aviation experts, including top military officers, Curtis reasoned that flight control methods were not only antiquated but also inhibited industrial development and even military planning. To meet future needs, Curtis recommended that Congress consolidate all federal aviation activities into a single organization with a "modern organizational structure." To begin the task, he suggested setting up a temporary agency for the purpose of testing and selecting new navigational aids and practices. This temporary agency—the Airways Modernization Board (AMB)—should include representatives from both the Departments of Defense and Commerce, and should be chaired by a widely respected aviation expert appointed by the president.[61]

Eisenhower sent Curtis's report to Congress along with a letter strongly supporting its recommendations. "This report sets forth the gravity of our present and anticipated air traffic problems," the president wrote. Eisenhower even attached a draft of legislation to the letter establishing the proposed AMB. Of the proposed legislation, the president said, "This measure will greatly expedite the improvement of air traffic control and air navigation and I therefore urge its early enactment."[62] To prod Congress in the desired direction, Eisenhower appointed one of the nation's most distinguished military officers, former Air Force General Elwood Quesada, as his new "Special Advisor for Aviation." The appointment suggested that Quesada had agreed to chair the proposed AMB and thus added another level of integrity to Eisenhower's actions. Quesada played a pioneering role in developing new principles of air combat during the war and continued to shape aviation policies after the war ended.[63]

Congressmen read these developments subjectively, however, in ways that validated their own needs and purposes. Some conservatives opposed the creation of the AMB on the grounds that the federal government had already established too many regulatory agencies. In the House of Representatives, for example, Republican Harold Gross of Iowa described the proposed board as "another layer of fat on Government." "I am opposed to creating any more new boards in the Government," Gross said emphatically. "I do not understand why," he added, "when we have a CAA and a CAB and all the rest of these boards and bureaus and commissions that we have in the Government, why we have to establish here another new one."[64]

Liberals in Congress expressed their own concerns about the legislation. Those who had often supported greater regulation of economic affairs saw the proposed AMB as merely a "makeshift solution" and therefore pushed to establish a larger, more permanent agency in the here and now. House Democrat John Moss of California summed up the group's sentiments. "I see nothing in the proposal now before us that does a single thing except to adopt a policy of studied procrastination," Moss said. "We need to designate one agency and give it authority to do the proper job. If we cannot do it in this session of the Congress, let us do it in the Congress that convenes in January."[65]

Despite their differences, lawmakers ultimately followed the president's lead. Oren Harris, who chaired the congressional hearings in Las Vegas,

brought Eisenhower's proposed bill before the House and supported it in the strongest of terms.

The nation's air control methods, the congressman stressed, had simply not kept pace with recent developments in aviation. "While the technological revolution in aircraft proceeded at a phenomenal rate, the system of controlling these aircraft in our airspace has lagged behind," Harris told his colleagues. Harris described the president's proposal as both practical and progressive. "This bill is a constructive step to meet a crying need with least cost to the taxpayers," he said.[66]

With bipartisan support and little delay, Congress passed the Airways Modernization Act, which Eisenhower signed on August 14, 1957. As its proponents intended, the act established the AMB with Elwood Quesada as its director and gave the new agency broad powers to introduce new methods of air control. The act stipulated that Congress, within three years, would establish a more permanent, independent agency capable of taking over all government responsibilities in aviation.[67]

A series of new midair collisions in the late 1950s added yet more momentum to this regulatory movement. On February 1, 1958, two large military planes collided over the suburban community of Norwalk southeast of downtown Los Angeles, killing forty-seven people on the planes and a civilian on the ground. One of the planes, a C-118 cargo craft, crashed and burst into flames behind the Los Angeles County Sheriff's Office. The other, a smaller P2V naval patrol plane, crash-landed two miles away, in a large pit near a clay factory. The collision scattered plane parts throughout the urban area, causing extensive property damage. The US Navy's investigation of the accident once again pointed to the limitations of air control methods. The pilots, the investigation suggested, had not seen each other in time to take evasive action. "This is not considered a failure in the sense of negligence or poor technique, but a failure created by human limitations," the navy concluded.[68]

Two subsequent collisions involving commercial aircraft kept the safety problem at the top of political agendas. On April 21, 1958, an air force fighter jet on a training mission slammed into a United DC-7 airliner with forty-seven people on board a few miles southwest of Las Vegas. The military plane, an F-100 Super Sabre, was in a diving pattern when it hit the slower-moving airliner at twenty-one thousand feet. The collision ripped off the

jet's right wing and part of its tail section. One of the jet's two pilots ejected from the crippled craft but apparently could not open his parachute, and the other went down with the plane. The airliner also suffered serious damage to its right wing and spiraled out of control. Witnesses saw black smoke and flames trailing behind the plane before it crashed violently into the dry desert ground. Military investigators blamed the collision on the nation's air traffic control methods. "The primary cause of this accident was the inadequacy of the present control system, which allows two or more aircraft to occupy the same airspace at the same time and relies solely on the ability of pilots to see and avoid one another," they concluded. CAB investigators also recognized the system's shortcomings but proceeded to criticize the military for conducting training exercises within the confines of an established airway. They also acknowledged that airline cockpits made it difficult for pilots to detect approaching aircraft.[69]

A month later, on May 20, another military jet collided with a small commercial airliner carrying eleven people. In this case, a pilot in the National Guard was flying a Lockheed T-33 Shooting Star eight thousand feet over northern Virginia when the right wing of his plane struck Capital Airlines Flight 300, which was descending into Baltimore, Maryland. The airliner's tail assembly broke away from the fuselage, and the plane went into a tailspin before crashing near Brunswick, Maryland, killing everyone on board. The pilot of the military craft ejected from his plane and lived through the ordeal, but his fellow guardsmen on board died in the collision. CAB investigators blamed the crash entirely on the jet pilot, whom they accused of failing to exercise "proper and adequate vigilance to see and avoid other traffic."[70] The back-to-back tragedies further justified the creation of a powerful new regulatory agency capable of overseeing both civil and military flights.

Only one day after the collision over Brunswick, Senator Mike Monroney of Oklahoma and Representative James Roosevelt of California, both Democrats, introduced a new bill to modernize air traffic control. The proposed Federal Aviation Act, as it was called, had been months in the making. Monroney described the bill perfectly: "It is the result of thorough hearings, long study, hard work, and the cooperative effort of every major segment of American aviation and every affected department or agency of the Federal Government."[71] The bill had emerged from a subcommittee of the Senate Commerce Committee that Monroney chaired. As the congressman ex-

plained at the time, airline companies, trade unions, and the armed forces had studied the bill and supported it.

The legislation was two hundred pages long and had fourteen major sections that were broken into subsections. In short, it called for the creation of a new federal regulatory agency with broad responsibilities. The Federal Aviation Agency, or FAA, as it would be called, would take control of both the CAA and AMB and assume CAB's responsibility for setting civil air traffic rules. But the agency would also have the power to restrict the military's use of airspace, to set new standards governing the construction and inspection of all civil aircraft, and to promote research and development in air safety. The legislation further authorized the FAA to set rules and regulations governing wages and working conditions of pilots, and protected the rights of pilots and other employees in aviation to join trade unions and bargain collectively.[72] President Eisenhower expressed full support for the bill. In a message delivered to Congress on July 13, Eisenhower urged Congress to pass it as soon as possible: "The concept of a unified Federal Aviation Agency charged with aviation facilities and air-traffic management functions now scattered throughout the Government has won widespread support from the Congress and among private groups concerned with aviation."[73]

Congress acted quickly. The Senate passed the Federal Aviation Act the day after receiving the president's message, and the House passed it a week later with only minor changes. In supporting the act, reliably conservative lawmakers acknowledged the tragic nature of recent midair collisions, drawing special attention to the Grand Canyon tragedy. As Republican Senator Wallace Bennett of Utah reminded the Senate, the shock of the 1956 disaster "has not been dimmed for those of us who live close to the scene."[74] House Republican Harry Haskell of Delaware pointed to the larger impact of midair collisions. "Since November 1949, 314 persons have lost their lives in 7 separate major mid-air collisions," Haskell told the House.[75] President Eisenhower signed the Federal Aviation Act on August 23, 1958, and appointed Elwood Quesada as the FAA's first administrator.

The FAA's basic structure took shape over the next few months. Quesada appointed the agency's first officers, who oversaw the establishment of separate bureaus to regulate flight standards, air traffic control practices, and control facilities. The officers worked from command centers in New Jersey, Oklahoma, and Washington, DC, where they began integrating all federal air responsibilities into their bureaus. Among other things, that meant in-

corporating the activities of the recently established National Aviation Facilities Experimental Center (NAFEC) in Atlantic City, where engineers and technicians tested new air safety and navigation devices. As quickly as possible, they worked to integrate radar surveillance, flight analyzers, and other new technologies into the air management system.[76]

The FAA became one of the nation's largest and most complex federal agencies. By 1960, it employed about thirty-five thousand people, including air safety inspectors, mechanics, clerks, congressional liaisons, and many other occupational groups. The majority of these employees worked in the agency's air control facilities, which consisted of 225 airport towers, 35 en route communication centers, and 425 ground radio stations. Many other employees worked in the field of research and development at the research center in Atlantic City. But the FAA also relied on the services of scientists and engineers at private research universities. It awarded large grants to the Massachusetts Institute of Technology, for example, whose faculty played a particularly important role in the FAA's research activities.

In only a few years, Director Quesada and his staff transformed the nation's air management system. The process involved integrating military radar technology into FAA control facilities, placing high-altitude airways under radar surveillance, and appropriating thousands of miles of military airspace for the purpose of expanding commercial airways. It also meant wresting radio frequencies away from the air force and private broadcasters to improve air-to-ground communications, as well as raising standards governing flight control operations, airline maintenance practices, and job training programs.[77] By the time Quesada retired in January 1961, the transformation was complete. Pilots no longer flew in uncontrolled airspace or relied primarily on their own eyes and instincts to avoid collisions, and air traffic controllers had abandoned old ways of tracking planes and communicating with pilots, airline companies, and one another. This was the new world of aviation, the "jet age."

In his groundbreaking study of domestic issues and policy making, the political theorist John Kingdon wrote, "Sometimes crises come along that simply bowl over everything standing in the way of prominence on the agenda."[78] The 1956 Grand Canyon disaster did just that. Covered by news reporters far and wide, the midair collision reflected poorly on almost every firm and agency in the field of aviation. Military officers, business leaders, and law-

makers alike realized that the accident threatened their long-term interests and consequently supported greater regulation of air travel. By the summer of 1958, after yet more lives had been lost in air accidents, even congressmen committed to small government and low taxes agreed on the need to establish a new regulatory agency with sprawling facilities and thousands of employees. Pilots and their union had long called for such reforms and made their own contributions to this safety movement. Support also came from teams of technical experts in public and private agencies, to whom policymakers looked for new ideas and guidance. Creating a new regulatory regime, however, ultimately hinged on the availability and deployment of new navigational technologies. Radar and radio surveillance technologies were therefore particularly important parts of this story.

Exploring this disaster has helped to highlight two basic premises of this book: that material growth is a two-sided phenomenon and that safety reforms were not simply heaven sent. But the exploration has also helped to cast contemporary workplaces in a new light. Commercial airliners were more than fast-moving objects filled with passengers. These winged structures were also workplaces where employees carried out a range of jobs that required real skills and talents. These spaces were modern in every sense of the word. They were designed in advanced research laboratories and manufactured in highly capitalized, technologically sophisticated factories. Once assembled, they could take to the sky like birds of prey, reach speeds that once seemed incredible, and return to earth with the precision of a guided missile. When these things crashed and burned, as they sometimes did, they underscored the perils of modern life. The midair collision over the Grand Canyon exposed modern perils like no previous air accident. In its wake, the days of flying through uncontrolled airspace went the way of barnstormers in biplanes.

4

Hospitals Collapse in Southern California!

Come, ye philosophers, who cry, "All's well,"
And contemplate this ruin of a world.
Behold these shreds and cinders.

Voltaire, "Poem on the Lisbon Disaster"

The earthquake hit just before sunrise, on February 9, 1971. Centered a few miles north of Los Angeles, it measured 6.6 on the Richter magnitude scale, making it the most powerful underground eruption to rock Southern California in nearly fifty years.[1] The eruption killed at least sixty-four people and injured more than twenty-five hundred. It took its heaviest toll in the San Fernando Valley, a sprawling suburban area near the epicenter of the great disturbance. In a few frightful seconds the ground movement wrecked large parts of the area's industry, infrastructure, and private property. More importantly, it wrecked two large hospitals, including a long-standing Veterans Administration (VA) facility where more than two-thirds of the victims perished. The movement also battered two other hospitals so badly they had to be evacuated. At a time when thousands of local residents needed immediate medical attention, the damage to these medical facilities seriously strained the community's ability to provide basic emergency services.

The impact of the 1971 San Fernando earthquake is revealing on multiple levels. To begin with, it underscores what scholars have long said about "natural" disasters: that many are "socially produced."[2] One of Europe's most influential philosophers, Jean-Jacques Rousseau, advanced that argument in the mid-eighteenth century, after a massive earthquake all but destroyed Lisbon, Portugal. In a letter to his intellectual nemesis, Voltaire, Rousseau

rejected the idea that Lisbon's ruins proved the absence of a higher providence, as the older Voltaire had recently suggested. To Rousseau, the devastation simply showed that the natural world was not kind to those who ignored its ways. "It was hardly nature that brought together twenty-thousand houses of six or seven stories," Rousseau wrote. "If the residents of this large city had been more evenly dispersed and less densely housed, the losses would have been fewer or perhaps none at all."[3] Though the Lisbon earthquake was far more destructive than the tremor that shook Los Angeles in 1971, Rousseau's message still rang true: urban developments had literally paved the way for a disaster. Had the quake occurred only a quarter century earlier, when the valley was still a thriving agricultural area, nature would have had far fewer structures to wreck.

This tragic event also shows how disasters have upended the lives of American workers—in this case, men and women in the expanding field of health care. Historians have only begun to explore the rise of modern health care facilities and the daily routines and challenges of their employees. Before the quake struck, the valley's hospitals resembled busy beehives. Administrators managed the day-to-day operation of the facilities; doctors and nurses interacted with patients; pharmacists oversaw the distribution of medications; and scores of clerical workers, ambulance drivers, and other employees attended to their duties. How did these health care workers react to the earthquake? What did the quake mean for their daily lives and routines? How did their employers help them cope with this disaster? These are the kinds of questions that inspired this study.[4]

The tragedy speaks to another matter of innermost concern: the close connection between disasters and public policies. In 1971, neither the City of Los Angeles nor the State of California had adequately addressed the earthquake problem. Seismic safety codes still varied from city to city, and only a few elected officials pushed to standardize or strengthen those codes. The number of seismic safety advocates expanded broadly after the 1971 quake. Public officials at all levels of government worked hand in hand with scientists and structural engineers to develop better, more comprehensive earthquake mitigation plans. This seismic safety movement advanced on several fronts, but ensuring the safety of hospitals received top priority. California strengthened building codes for medical facilities in the state through the Hospital Seismic Safety Act, and the VA adopted similar regulations for its own facilities. The VA also closed several hospitals in quake-prone regions

of the country, including its wrecked facility in San Fernando Valley. On the one-year anniversary of the quake, the agency awarded the land on which its old hospital had stood to Los Angeles County, which turned the site into a regional park.[5]

These changes did not happen overnight. They materialized gradually, in bureaucratic arenas governed by procedures and rules, where some individuals and groups held more power and influence than others. In some ways, this story of policy change resembled a classical tragedy in which a community struggled to find stability in an unstable, anxiety-ridden setting. The story had a large cast of characters that performed before the general public, who followed the action through news reports, public hearings, and official statements. The quake's victims took center stage and generated a great deal of sympathy and support, but the story's protagonists were mostly elected officials who were well positioned in the institutional and political environment. Technical experts played a crucial role behind the scenes, advising the officials on all matters related to seismic safety. "Mother Nature" may have been the chief antagonist, but urban developers and builders bolstered the threat that she posed. Our plot reached its climax with the passage of new seismic safety laws, which finally restored a sense of stability to the stage. This drama's denouement stretched on longer than expected, as the protagonists struggled to implement safety reforms.[6]

The San Fernando earthquake took place against a backdrop of head-spinning urban growth. Wartime defense spending created an abundance of job opportunities in the West, luring tens of thousands of Americans to the region. California's population grew at a faster clip than any state in the nation. A million people moved to California during the war, increasing the state's population by roughly 15 percent. The vast majority of the newcomers settled in urban areas, adding to the population of some cities by more than 25 percent. On closer inspection, most of this growth was centered in Southern California, in the metropolitan areas of Los Angeles, Long Beach, and San Diego. There, defense spending spawned the rise of giant shipyards, aircraft factories, oil refineries, steel plants, and other large enterprises. No other part of the country experienced such dramatic growth.[7]

The influx of capital and people into Southern California remained heady throughout the postwar years. Many of the newcomers were among the ten million servicemen or women who were stationed in the area and liked what

they saw. Others were among the millions of Americans who traveled there after the war and decided to stay. Between 1945 and 1965, while the population of several northeastern and midwestern states rose by less than 10 percent, Southern California's population more than doubled, jumping from roughly five to eleven million. By 1970, more than seven million people lived in the county of Los Angeles alone.[8]

Southern California's surging population had enormous consequences. Among other things, it strained public services. In Los Angeles, population growth overburdened public utility facilities, water storage systems, and transportation networks. Public schools became overcrowded; by midcentury, because of room shortages, thirty thousand children in the "City of Angels" attended school in shifts. These problems partly explained a remarkable construction boom on the outskirts of Los Angeles and San Diego, and eastward toward the hills of San Bernardino. By 1963, home construction in Southern California accounted for nearly a quarter of the nation's homebuilding activity. As news reporters liked to point out, residential neighborhoods ripened faster than local orange orchards and olive groves. They might have added that developers had already plowed over much of the area's farmland to make room for new homes.[9]

The transformation of San Fernando Valley in these years mirrored the broader patterns. Covering more than 250 square miles between two parallel mountain ranges, the triangular-shaped valley quickly transitioned from an agricultural mecca to a tangled web of diverse neighborhoods. Between 1945 and 1960 the valley's population quintupled, jumping from 170,000 to 850,000. Los Angeles promptly annexed most of the valley and thereby allowed the new communities to contract the city for a range of municipal services. By 1970 the valley's population topped one million, making the area more densely inhabited than most Sunbelt cities, including New Orleans, Phoenix, San Diego, and even Houston. Unlike those cities, however, the valley was quintessential suburbia. Most of its working residents commuted daily into the more populous Los Angeles Basin, driving across eastern ridgelines of the Santa Monica Mountains and continuing southward to the downtown area.[10]

The daily commutes took many residents along the foothills of the San Gabriel Mountains, which loomed ominously over the valley's northeastern neighborhoods. The Mountains were sixty miles long and twenty miles wide, forming a rugged, imposing barrier that separated the Los Angeles area from

the sparsely populated Mojave Desert. They had their origins in ancient volcanic eruptions along the mammoth San Andreas Fault that stretched from Northern to Southern California and through the mountains' eastern rim. Geologists warned repeatedly that the San Gabriel Mountains were especially unstable. The mountains' steep, rocky slopes concealed several major breaks in rock formations, and these "faults" moved horizontally or vertically in response to seismic activity deep beneath the earth's crust. At times the movement had been so ferocious it changed the entire landscape.

The ground motion that triggered the 1971 quake occurred along the mountains' western rim, near the valley's northeastern margins. At 6:01 a.m. that fateful February morning, the ground fractured about eight miles beneath the town of Newhall, an emerging community just beyond the valley's boundaries. The rupture sent shock waves radiating outward at speeds of thirty thousand to thirty-five thousand feet per second. When the waves reached the valley's soft, water-saturated clay, they spiraled downward, churning up dirt and rocks like an army of giant, superpowered bulldozers. Stopping abruptly, the waves released enough energy to jolt millions of Southern Californians out of bed. People more than two hundred miles from the epicenter felt the earth move that day, in places as far away as Las Vegas, Nevada, and Tijuana, Mexico. Seismologists compared the quake's force to atomic bombs dropped on Japan near the end of the war.[11]

Whatever the appropriate comparisons in physical force, the ground movement resulted in widespread economic and social disruption. In greater Los Angeles, it damaged twenty-one thousand single-family homes, seventeen hundred mobile homes, and a hundred apartment buildings, leaving some fifteen hundred families in need of temporary housing.[12] It damaged another five hundred commercial buildings, most of which were so badly shaken they were eventually deemed unsafe. Unemployment consequently spiked in the area. More than thirteen hundred local residents filed unemployment claims within two weeks of the quake, and about three thousand others filed claims soon thereafter.[13] The tremor also damaged more than a hundred schools in the area, many of which had to close for repairs. Nearly ten thousand children transferred to new schools after the quake, some on a permanent basis.[14]

The shaking in the valley's northeastern foothills was almost unimaginable, even by California standards. To get a sense of the quake's power, consider how hard it shook a large Catholic seminary in the area—Our Lady

Queen of Angels—and the adjacent San Fernando Mission, California's oldest and perhaps most revered Spanish mission. There were two hundred boys in the seminary when the shaking started, and Reverend Francis Weber had just begun his daily work routine. The priest later recalled the moment: "The lights went out and the whole building began rolling," he explained. "Every single one of the more than 5,000 books lining the walls began flying through the air; shelves collapsed; pictures catapulted from the walls; glass crunched; file drawers careened off their tracks and dresser cabinets sprang open." When the shaking finally stopped, Weber rose to his feet, grabbed a flashlight, and ran through the dormitories, instructing frightened students to gather near the mission's cemetery. There, the priest and students saw that the quake had not only damaged the seminary but also wrecked the nearby mission. "A rain of plaster and adobe buried the whole interior of the church beneath several inches of crumbled debris," Weber lamented. Among the debris were pieces of the mission's cherished crucifix, statues, and other treasures.[15]

Like the historic eighteenth-century Spanish missions that dotted California's coast, many homes and businesses in the state were not built to withstand significant seismic forces. Seismic safety standards in California were still relatively weak and varied from place to place. The great earthquake that ravaged San Francisco in 1906 proved that simple brick buildings were less resistant to violent ground movement than were wooden structures and concrete buildings reinforced with steel wires and bars. Such lessons, however, had had little effect on state building codes. San Francisco and other cities continued to allow the construction of unreinforced brick buildings well into the twentieth century.[16] Not until 1925, when an early-morning tremor rocked the coastal city of Santa Barbara, did municipalities begin restricting the use of brick masonry. The Santa Barbara quake destroyed a four-story brick hotel and several other buildings in the city's business district, killing twelve people. Investigations of the damage revealed that properly braced concrete buildings withstood the quake far better than older brick buildings and that some of the latter structures showed signs of poor workmanship. In response, city leaders began revising local building codes, placing new restrictions on the use of brick construction.[17]

It took a much larger disaster to produce statewide reforms. On March 10, 1933, an earthquake centered off the port of Long Beach brought down more

than a hundred buildings in the greater Los Angeles area, including more than a dozen public schools. Fortunately, the quake struck about five o'clock in the evening, shortly after schools and businesses had closed. But it nonetheless killed about 120 people and injured nearly 5,000. It also left thousands of other people homeless, including entire families that had already been hurt badly by the Great Depression. In the 1930s the nation had no federal disaster-response agency, and thus the quake's victims had to rely on humanitarian organizations for food, clothing, and other forms of assistance. The Red Cross fed about eighty thousand people a day after the quake, which helped convince newly elected President Franklin D. Roosevelt to send ashore some four thousand servicemen from nearby naval vessels to support relief efforts.[18]

News reports described the 1933 quake as a natural disaster, but the Los Angeles County coroner's office concluded that its impact was largely manmade. The coroner, Frank Nance, was a well-known figure in Southern California. In the 1920s and early 1930s, Nance investigated several high-profile Hollywood murders and suicides, and came to be known as the "Coroner to the Stars." The 1933 disaster had created new mysteries for Nance to resolve, including the curious deaths of two teenage girls. Nance's inquest blamed the girls' deaths on shoddy practices in the local construction business. As the coroner's report noted, many structures in the Long Beach area had been built hastily, cheaply, and with little concern for public safety: "Stability of construction was often sacrificed for architectural effects or for purposes of utility or convenience, or with an eye to economy." The report also lambasted public officials for not properly regulating construction practices. Those officials, it said, had shown "indifference to adequate building practices."[19]

In moves that underscored the reactive nature of reform movements, California legislators passed two bills after the Long Beach quake. The first, presented by Republican Assemblyman C. Don Field, transferred regulatory power over public school construction from municipal governments to the state. In effect, the bill required all new schools to be designed and constructed by state-licensed architects and engineers in ways that integrated recent advances in earthquake engineering. It also made structural engineers within the state's Division of Architecture responsible for inspecting and approving their work.[20] The second bill, introduced by Assemblyman Harry Riley of Long Beach, also a Republican, required all local governments

to establish building departments to oversee construction practices in their jurisdiction. The bills essentially banned the construction of unreinforced brick buildings and thus encouraged the rise of new steel-reinforced concrete structures that could slightly bend and sway without buckling. Representatives from the building industry saw both bills as onerous and expensive, but Republican Governor "Sunny Jim" Rolph Jr. signed them into law.[21]

Public officials hailed these two acts as milestones in the history of public safety, but the laws had obvious shortcomings. To begin with, the seismic safety standards they set for new buildings were too low to resist the lateral forces produced by unusually powerful earthquakes. The acts also permitted the continued use of all existing brick buildings, including brick schools. Moreover, they failed to provide enough financial support to enforce new building codes. Most local governments lacked the funds to employ building inspectors with expertise in seismic safety, and even successful architects and engineers did not fully understand how buildings responded to seismic forces. Though legislators eventually amended the laws to encourage the reinforcement of older schools and other large structures, builders continued to ignore them.[22]

Underground eruptions unnerved Southern Californians throughout the war and early postwar years, but the tremors had little effect on seismic safety standards. In 1941, an earthquake again damaged structures in downtown Santa Barbara, some of which had not been repaired properly after the city's 1925 temblor. A few months later, another ground-shaker roared through southeastern neighborhoods of Los Angeles, damaging about fifty buildings in the communities of Torrance and Gardena and knocking out local gas and water services. A much more powerful quake in 1952 killed nine people in the railroad town of Tehachapi, about a hundred miles north of Los Angeles, and left the town's railroad tracks "twisted like a licorice stick," as one reporter put it.[23]

Despite these quakes, there was little support for seismic safety reform before March 27, 1964, when an earthquake in southern Alaska raised questions about the safety of everyone along the West Coast. With a magnitude of 8.5 (and perhaps more), the quake disrupted nearly two hundred thousand square miles of ground, an area larger than the entire state of California. The titanic upheaval destroyed several buildings in Anchorage, more than eighty miles from its epicenter, and generated a series of tsunamis that threatened the coast of Oregon and Northern California. The Great Alaskan

Earthquake, as it came to be known, killed only 131 people. But it nonetheless had major implications in the realm of earthquake research and engineering. In its wake, scientists and engineers reached new conclusions about the value of soil sampling, fault mapping, and ground monitoring. Agencies like the US Geological Survey and California's Division of Mines and Geology became strong advocates for seismic safety reforms and started pressuring public leaders to develop earthquake mitigation programs. In cities like San Francisco and Los Angeles, where earthquakes posed the greatest threat to public safety, the news media paid close attention to the work of these agencies and to the warnings of their spokesmen.[24]

If the Alaskan earthquake laid the foundations for a reform movement, a subsequent quake reinforced them. On April 8, 1968, rock formations fractured beneath a sparsely populated area about a hundred miles east of San Diego, along the well-mapped San Jacinto Fault near Anza-Borrego State Park. The quake was the strongest to strike California in fifteen years. It sent giant boulders crashing down mountainsides, ramming vehicles and blocking roads. Though it claimed no lives, the quake severed power lines in San Diego and cracked ceilings and walls as far away as Los Angeles. More importantly, the quake displaced ground along three additional faults that ran through Southern California, raising the question of whether a future tremor might trigger several simultaneous earthquakes.[25]

The specter of such a calamity prompted Los Angeles County Supervisor Kenneth Hahn to call for a complete review of all local seismic safety ordinances. Hahn's emergence as a seismic safety advocate represented a significant development. He was one of Southern California's most recognized and respected civic leaders. A former naval officer in World War II, Hahn had held public office in Los Angeles since 1947. As county supervisor, he played a vital role in bringing major league baseball to Los Angeles and helped improve the city's transit system and medical facilities.[26] The issue of seismic safety was also a personal matter to Hahn. He was a survivor of the 1933 Long Beach quake and thus had seen how devastating quakes could be. In the wake of the Anza-Borrego tremor, Hahn reminded his Board of Supervisors that earthquakes posed a grave risk to the entire state. As he explained to the board in a meeting at County Hall of Administration, prominent scientists and engineers had recently warned Governor Ronald Reagan of an impending disaster. "These warnings cannot be ignored," Hahn told the

board. "Action must be taken at both the local and state level to eliminate potential hazards." The city's new downtown skyscrapers, he added for effect, could "topple like dominoes in a disastrous earthquake."[27]

At Hahn's urging, the board formed an advisory committee to seek advice from earthquake specialists at the California Institute of Technology (Caltech). Located in nearby Pasadena, Caltech had been at the forefront of seismological studies since the early 1930s. During the war and early postwar years the federal government poured millions of dollars into the university's seismology programs, not just to minimize seismic dangers but also to find better ways of detecting underground nuclear tests. The university grew even richer after the 1952 Kern County quake, when it convinced local railroad companies, insurance agencies, and other large corporations to invest in the development of earthquake-resistant structures. By the time Hahn's advisory committee reached out to Caltech for advice, the university employed some of the world's most eminent seismologists and engineers, including Charles Richter, who had helped to develop the method of measuring the power of earthquakes that bears his name—the Richter magnitude scale. Richter advocated for stronger seismic standards in California since joining Caltech's faculty in the late 1930s and had recently called for renovating structures that might collapse during a quake. Though recently retired, Richter continued to promote earthquake preparedness through public interviews and professional talks. His willingness to work with Hahn's committee was no small matter—it added instant credibility to the committee's mission and work.[28]

With the help of Richter and other Caltech faculty, the advisory committee drew up a long list of seismic safety recommendations. One of these recommendations called for the County Engineer's Office to map all faults in the Los Angeles area, and it proposed a prohibition of new construction along those lines. Another suggested that owners of existing properties along the lines should meet a special set of building code requirements, and still another said that real estate companies selling structures near faults should divulge geological conditions beneath their properties. The committee also recommended that companies involved in the distribution of gas, oil, or chemicals should install newly developed shutoff valves on all pipes that crossed faults. The valves, as experts explained, activated automatically if the ground suddenly shifted. Despite vociferous protests from the con-

struction and real estate industries, Hahn and his board accepted most of the committee's recommendations and were in the process of adopting them when the San Fernando quake hit.[29]

The tectonic shock of 1971 brought the need for new seismic safety regulations into sharp relief. Along the northeastern margins of the San Fernando Valley, old communities like Sylmar and Pacoima resembled war zones.[30] A thriving business district along the area's main thoroughfare was suddenly in ruins. A hardware store, barbershop, sporting goods outlets, and other small firms had partially collapsed, and fires burned through a shopping center for nearly two hours.[31] Public utility facilities were in shambles, too. General Telephone Company lost its main "switching center" and many of its telephone poles and lines, and the Los Angeles Department of Water and Power lost an electricity-generating plant as well as water storage tanks. The Southern California Gas Company reported 450 breaks in its underground pipelines, several of which had sparked fires. The damage left thousands of valley residents without electricity, gas, heat, and running water.[32]

The damage to the Van Norman Dam, located north of the valley, posed the most urgent problem. The 1,100-foot dam shored up a water reservoir containing more than half of the Los Angeles area's water supply, or roughly 3.6 billion gallons of water. Built in 1915, the dam's concrete rim had previously stood thirty-six feet above the water level, but large portions of the rim crumbled during the quake, leaving it only six feet above the water. After inspecting the damage, engineers concluded that the quake's frequent aftershocks could bring down the entire structure, possibly killing tens of thousands of valley residents. The image of a huge tidal wave washing across the valley convinced Los Angeles Mayor Sam Yorty to order the immediate evacuation of all residents living directly below the dam, or roughly eighty thousand people.[33]

The mayor's evacuation order created one of the worst traffic jams in Southern California's history. Several of the valley's bridges and overpasses collapsed during the quake, and almost all major surface streets showed some signs of damage. Most of the valley's traffic lights were inoperable, too, and numerous street signs had been uprooted or badly twisted. State and city agencies did all they could to facilitate the evacuation process. State Highway Patrol officers set up elaborate detours around torn-up roads and freeways, and radio newscasts informed drivers of major traffic accidents

and tie-ups. Yet getting out of the valley proved so difficult that some drivers opted to return to their homes.[34]

Meanwhile, city engineers worked furiously to draw down the reservoir's water level. They received vital support from the US Army Corps of Engineers' emergency operations center in Los Angeles. When the corps learned of the dam's vulnerability, it rushed several large pumping machines and a team of corpsmen to the site. By nightfall, the hastily organized task force had cut openings in the dam's earthen base and begun transferring water from the reservoir into the Los Angeles River, which meandered through downtown Los Angeles before emptying into the Pacific Ocean. For the next three days, the men pumped 250,000 gallons of water per hour out of the reservoir, until its water level dropped by ten feet. At that point, the mayor permitted weary residents to return to their homes.[35]

Driving along the valley's eastern edges, many of those residents no doubt watched the brush-covered foothills of Sylmar, where they heard that a long-standing Veterans Administration Hospital had collapsed. The VA facility opened in 1926 as a tuberculosis treatment center, and it covered nearly a hundred acres of the gently sloping terrain. The facility originally consisted of six wards for patients and about ten other structures, including an administrative building, nurses' quarters, a warehouse, and a fire station. It expanded noticeably at midcentury, however. As medical science developed new vaccines for tuberculosis and the valley's population surged, the place evolved into a general hospital for veterans. By 1971, it had two more buildings designed to house patients as well as a new research laboratory and a boiler plant, paint shop, and chapel. The maze of structures accommodated 420 patients and 560 employees.[36]

Though the hospital's newer structures survived the quake with relatively little damage, four of its older buildings completely collapsed, including a six-story structure filled with cardiac and respiratory patients. According to eyewitnesses, the building folded "like a house of cards." The hospital's chaplain, Reverend Edward McHugh, had been walking alongside the building when it buckled. "I got thrown to the ground about 20 yards from the building," the chaplain told a reporter. "When I looked up, in that same instance, the building fell. Just like that, it fell, in a single second."[37] The collapse of the structure left more than eighty people dead or entombed in rubble, including more than a dozen employees.

Survivors of the quake worked frantically to help those who were not so

Aerial view of the quake-stricken Veterans Hospital in Sylmar, California, February 9, 1971. The quake killed forty-six people at the hospital, including ten of the facility's employees. Courtesy of the Los Angeles Fire Department and the University of California Library Special Collections

fortunate. Some dug through rubble in search of survivors, listening carefully for cries of help. Others carried the injured to makeshift medical stations at the site, where doctors and nurses cleaned wounds and calmed nerves. Several surviving employees tried to establish contact with the outside community. Without functioning telephones or radio equipment, staff members rushed into the surrounding residential area to seek help only to learn that no one in the area had telephone services. A few security guards

ran all the way to the VA hospital in Sepulveda, located about ten miles south of Sylmar, where they pleaded for additional ambulances and medical staff.[38]

The person who directed these rescue activities, chief hospital administrator Joseph Heavey, later discussed the challenges he faced that morning. Heavey lived with his wife and children at the hospital and was still asleep in his quarters when the earth roared. "For some fifty seconds my family and I held onto our mattresses, covered our heads with pillows, and prayed," he explained. "I could hear glass and dishes breaking throughout my quarters on the hospital grounds, parts of the ceiling fell in, parts of the wall fell toward the lower portion of my bed. It was a terrifying fifty seconds." When the shaking stopped, Heavey bolted from his bed, dressed, and tried to contact the hospital's telephone operator. "There was no response, and it was obvious the telephone lines, as well as all electricity, were out." The administrator then used a portable, two-way walkie-talkie to reach the operator, who told him buildings had collapsed. After making sure his family was safe, Heavy rushed toward the hospital's main structures, where he saw the most horrifying sight. "Buildings 1 and 2 had reduced themselves from fine structural buildings for patients to rubble," he recalled. "Wards 4, 5, 6, and 7 and the Dietetic Department were a part of this complex."[39]

The administrator's training in the field of civil defense probably saved a few lives that morning. Taking a "bullhorn" from one of the hospital's own fire trucks, Heavey began marshaling physicians and nurses to the collapsed buildings, and instructed them to help the injured. "Our VA physicians gave hypodermic injections, supplied oxygen and administered to patients who were in the most dire need."[40] Heavey immediately ordered the evacuation of patients from buildings that were still intact, and he established triage areas far away from the buildings to avoid dangers posed by aftershocks. "The patients were carried out on mattresses, wheel chairs, and anything available making it comfortable for them." As doctors and nurses attended to patients, Heavey sent guards into the community to find assistance. As he later explained, it took more than an hour for emergency vehicles to arrive at the hospital. "For approximately one and one-half hours after the first destructive jolt, no one knew what was happening on the San Fernando VA Hospital grounds except those of us who watched."[41]

The absence of police and rescue vehicles at the hospital reflected the chaos and confusion in the valley that morning. One of the hospital's guards

apparently flagged down a police car as he ran for help, but when he told the officer about the crisis, the officer assumed the guard was reporting the collapse of structures at the nearby Olive View Hospital, where police cars were already present. As a different police officer, Al Fried, later told the press, "It was another 30 to 45 minutes before we realized there was a problem at the Veterans Hospital." By then, traffic jammed the road, and even police cars with sirens blaring had trouble reaching the devastated hospital. "Traffic was so tied up that it took another 25 to 30 minutes just to get there," Fried explained.[42]

The County Fire Department reached the hospital around 7:30 in the morning, just as police cars began to arrive. One of the department's helicopter pilots reported seeing structural damage to the hospital while searching for fires in the area. Fire Chief Raymond Hill then sent a second helicopter to the facility for closer inspection, and its crew quickly realized the scope of the calamity. "One of my deputies landed there and was the first man in there," Hill later told county officials. "He talked to a doctor, and they said they had been unable to get any kind of communication outside." Hill then sent a large task force to the hospital. "We immediately put a 160-man force in there, including some nine ambulances."[43] The crew had both the skills and the tools to rescue and treat the injured, including paramedics with specialized medical gear designed to save patients in cardiac arrest. By noon, the crew had helped evacuate more than three hundred people from the grounds, many of whom had been lifted off the grounds by helicopters. Others left in commandeered school buses, delivery trucks, and smaller vehicles.[44]

To find survivors deep beneath the rubble, the fire department turned to the ubiquitous Corps of Engineers. According to one of the corps' officers in Los Angeles, Colonel Robert J. Malley, the agency received a call from firemen at 7:45 that morning requesting help in freeing survivors. "We knew the seriousness of the situation as we had been viewing it on a television screen in the Operations Center [and] we immediately began action," Malley said.[45] Two of the agency's engineers rushed to the hospital, and its chief officers soon followed. "Shortly after nine that morning, members of my staff and I departed for the disaster area by helicopter," the colonel explained. "Reconnoitering the area rapidly, we met with personnel from the Los Angeles County Fire Department who were directing the emergency operations, and placed our personnel, equipment, and other resources at their complete disposal."[46] The resources included thirty-five corpsmen and several

Rescue workers pull victims out of the ruined Veterans Hospital in Sylmar, California, February 9, 1971. Courtesy of the Los Angeles Fire Department and the University of California Library Special Collections

giant cranes in addition to trucks, jackhammers, concrete saws, and boxes of small tools.

The search and rescue operations continued around the clock for four days, during which time rescue workers pulled thirty-eight survivors from the ruins. The last survivor was sixty-eight-year-old Frank Carbonara, a baker who spent two and a half days gasping for air under slabs of concrete in the hospital's cellar kitchen. Carbonara had arrived for work only minutes before the quake destroyed his workplace. When the floor tilted and ceiling

Restoring order at Sylmar's Veterans Hospital, February 9, 1971. Courtesy of the Los Angeles Fire Department and the University of California Library Special Collections

tiles started to fall, the baker slipped under a nearby kitchen sink. "I thought I was dead," he said. "I closed my eyes and waited, said a prayer God would forgive my sins."[47] Wedged deep beneath the wreckage in total darkness, Carbonara reflected nostalgically on life's small pleasures. "I thought of my wife's face in the morning and the way it felt to drive my first car. I thought of sky, and trees, and the way bread smells fresh from the oven." After lying motionless for fifty-eight hours, he heard a voice ring out. "And then I heard noises," he remembered. "They had found me. Who knows how or why?"[48]

Miraculously, Carbonara only had a broken wrist and hand, and was soon home with his family. By then, rescue workers had discovered forty-six dead bodies at the hospital, including those of ten employees, seven of whom were women. Another twenty people suffered severe injuries.[49]

The collapse of the VA hospital was part of a larger crisis in the local medical community. The quake damaged more than a dozen other hospitals in the greater Los Angeles area, including three of the valley's largest and newest facilities. It also cracked walls and ceilings at many smaller medical center buildings, especially those closest to the epicenter. Even facilities with relatively little structural damage experienced power outages and mechanical failures, making it difficult for their doctors and nurses to provide basic medical care.

The county-owned Olive View Hospital in Sylmar suffered an especially cruel blow. Like the VA hospital, Olive View had opened in the 1920s as a tuberculosis sanatorium and evolved into a sprawling, general care facility. In the 1960s, doctors performed one of California's first open-heart surgeries in the hospital, and the University of California, Los Angeles, began setting up its own medical programs there. By the 1970s, Olive View served residents well beyond the northern boundaries of the San Fernando Valley. It employed some fifteen hundred people and housed more than five hundred patients. Yet nearly half of the hospital's fifty structures were more than thirty-five years old, making everyone within them vulnerable to seismic shocks.[50]

Surprisingly, however, Olive View's newest structures suffered the greatest damage. The quake almost brought down the facility's new six-story Medical Treatment Center. The cross-shaped building's support columns cracked, twisted, and lurched upward during the quake, and three huge stair towers peeled away from the building and crashed to the ground like megaliths. Several concrete slabs also broke off from the main structure, and a carport above the emergency entrance caved in, too. Inside the building, beds, chairs, and tables ricocheted off walls; lights went out, windows shattered, and elevator doors jammed. Despite the heaving and flailing, only three people lost their lives. An ambulance driver perished beneath the collapsed carport, and two patients died when their electrical pressure-breathing equipment failed.[51]

The damage was far worse at Olive View's new two-story psychiatric

Sylmar's Olive View Hospital, February 1971. The earthquake brought down three of the hospital's giant stair towers, a carport, and a two-story psychiatric unit. It left other parts of the facility beyond repair. Courtesy of the Los Angeles Department of Water and Power and the University of California Library Special Collections

building. Shaped like a T, the concrete structure's second level housed patients, and its lower level accommodated administrative offices, examination rooms, and a large dining area. When the ground started shaking, the building's upper level "pancaked," crushing everything below it. As luck would have it, the quake struck before the first floor's administrative staff arrived for work. Even the floor's security guard was outside the building. But more than two hundred patients and employees from the higher floor fell like rag dolls into the dust and debris, and many of them suffered serious injuries. Almost all needed immediate medical attention, including many patients with acute mental health problems. Some of those patients picked themselves up after falling and wandered through the rubble dazed and con-

fused, looking more like zombies in a horror movie than real people with real problems.[52]

Victor Winfrey, one of several medical assistants in the so-called psych unit that morning, later described an unforgettable experience. Winfrey was attending to a patient chained to a bed when everything started shaking and then went dark. "I knew it was a quake and I wanted out, but I had a patient in restraints and I couldn't see." As the shaking intensified, Winfrey heard the sounds of "swaying, snapping construction." At that point, something even more unexpected happened. "The bed bounced up and hit me twice." The jolt temporarily stunned Winfrey, but he quickly collected himself and stumbled toward a flashlight in the hallway. "I unlocked the patient and dragged him after me," he explained. "We had to get out."[53] The nurse and his patient quickly reached a hallway door that was "jammed shut." At that point, Winfrey remembered someone picking up a chair and heaving it toward a window, but the windows were shatterproof and the chair bounced backward. Winfrey then noticed a nearby open door and darted through it with his patient in tow. To his great surprise, Winfrey found himself at ground level. "When the cool air hit my face I was astonished," he explained. "We were no longer on the second floor, but the first. What had been the first floor, with clinics and dining rooms, was only rubble beneath us."[54]

Nurses at two nearby hospitals also had extraordinary experiences. At Holy Cross Hospital, a ten-year-old, 260-bed facility that employed five hundred people, Joan Farias was preparing patients for early-morning surgery when the quake knocked her off her feet. Farias remembered hearing a loud boom before falling and then felt the hospital sway violently from side to side. When she looked up, televisions fell from their stands, beds bounced up and down, and medical tools flew through the air. In the darkness, Farias regained her footing and tried to calm patients, many of whom had jumped from their beds and rushed from the room. "The earthquake produced instant ambulation," she explained. Other nurses ran through her floor telling everyone to evacuate the building, but that proved difficult. The quake shut down the hospital's elevators and lighting system, and left its stairwells cluttered with debris. But Farias and her coworkers navigated the clutter and led patients to safety. "We had one line going up the stairway to evacuate patients unable to walk, and one coming down," she explained.[55]

The evacuation of the Pacoima Lutheran Hospital, a 110-bed facility that had also opened a decade earlier, was no less difficult. The hospital's director

of nursing, Bethene Peterson, discussed the challenge. "We had no phones, we couldn't get doctors, and over half of our help lived in areas from which it was impossible to get in because of blocked roads," Peterson remembered. The quake lifted the entire facility about a foot off the ground and dropped it like a rock. Broken glass littered floors, beds obstructed doorways, and elevators stalled. Patients awakened, jumped out of beds, and rushed toward the exits. Shards of glass and dancing furniture cut and bruised many of those patients as well as several nurses, compounding the existing problems. As though the scene needed more drama, an explosion at a nearby industrial plant sent carloads of critically injured workers to the hospital, and administrators like Peterson had to turn them away. Rooms of newborn infants also had to be evacuated. Peterson remembered seeing nurses bouncing cribs filled with babies down stairwells. "Pediatric nurses put five children to a crib, and down the stairs they came."[56]

The blow to valley hospitals forced the transfer of nearly twelve hundred sick or injured patients to other medical facilities—a daunting task. Major thoroughfares were not only inaccessible or heavily congested, but emergency vehicles were also in short supply, and many nearby facilities were too badly damaged to admit new patients. The quake immobilized elevators and medical equipment at the Kaiser Hospital in Panorama City, for example, and shook the Indian Hills Medical Center in Granada Hills so badly that the facility had to evacuate its own patients. Most of the northeast valley's displaced patients had to be transported many miles to receive proper medical care. Olive View transferred its mental patients to Camarillo State Hospital, located fifty miles northwest of Sylmar, for example, and sent its surgical cases to the County–University of Southern California (USC) Medical Center in downtown Los Angeles. The hospital moved most of its other patients to rehabilitation centers in the cities of Long Beach and Downey, also about fifty miles away.[57]

The quake left large medical staffs with no place to work. Though Holy Cross and Pacoima Lutheran soon reopened on a limited basis and brought their employees back to work, Olive View and the VA hospital were damaged beyond repair.[58] Los Angeles County temporarily kept about 200 of its Olive View employees at the hospital to perform custodial, maintenance, and clerical duties, but it eventually assigned the entire workforce to other medical facilities in the greater metropolitan area. It placed 375 employees in the Downey facility, 350 at the USC Medical Center, and 60 in the Long Beach

hospital. The rest went to county facilities in other parts of the metropolis. For almost all of these employees, the transfers meant longer commutes, different work routines, and new coworkers and supervisors.[59]

As one of the nation's largest employers, the VA was able to place most of its Sylmar employees in similar jobs at their existing pay. The agency kept about 450 of the employees in Southern California, many of whom transferred to one of three VA hospitals in the greater Los Angeles area. It sent another 50 employees out of the region and into more than thirty different VA facilities in other states. Another dozen or so took jobs in other federal agencies after the quake. Whichever path they took, the displaced employees had to adjust to new work environments, new commutes, and new daily routines.[60]

The San Fernando earthquake disrupted life for almost everyone in Los Angeles, but it left residents in the area of heaviest shaking in dire straits. Two weeks after the quake wrecked their hospitals, Sylmar residents were still without running water, telephone services, gas, and heat. With thousands of homes in disrepair, families were still camped out on lawns in crude tents, truck beds, and even horse trailers. Grocery stores in the community remained closed, and supplies like flashlights, batteries, and can openers were hard to find. Several schools canceled classes because of building damage, and attendance at all schools fell sharply. Many lost and hungry dogs were roaming through neighborhoods, some in frightening packs. These and other lingering problems prompted about fifteen hundred residents to hold a mass rally at a community baseball field, where they derided government officials for seemingly ignoring them. "They don't even know where Sylmar is," one of the protestors told reporters. Although the rally brought representatives from both the governor's office and federal agencies to Sylmar, it did little to speed recovery efforts. As one reporter at the rally noted, "Sylmar will not be rebuilt easily or soon."[61]

The Sylmar rally reflected anger over the slow pace of recovery, but it also revealed deep concerns about public safety. Even before the rally began, public officials had to respond to fears of another deadly quake. County Supervisor Hahn had revved up the long-running seismic safety movement by personally inspecting the VA and Olive View hospitals after the quake and by stating publicly that he intended to strengthen local building codes. To put the plan in motion, Hahn established a blue-ribbon panel to study

the 1971 quake and help develop the "best possible" codes.[62] He convinced Caltech's president, Harold Brown, to chair the panel. This Earthquake Commission, as it was called, included some of the nation's most distinguished scientists and engineers. Brown himself was an eminent nuclear physicist who had previously directed national defense research projects.[63]

The commission carried out its work methodically. Much of it was done in biweekly meetings in the county's new downtown Hall of Administration, where experts came to discuss the quake's severity and consider new seismic safety proposals. At one of the first meetings, a prominent Caltech geologist, Clarence Allen, explained how the 1971 quake had shifted the ground along faults. Though Allen applauded the county's efforts to restrict construction on these lines, he said much more should be done. "[We] have got to be better prepared next time it occurs." To begin with, Allen urged the county to create especially stringent building rules for hospitals and other medical facilities. "Our standards for building hospitals must be higher than our standards for other structures," he said.[64] Caltech engineering specialist George Housner agreed. "The lesson is clear that certain structures should be designed to be stronger than the minimum requirements of the building code," Housner told the commission.[65]

Supervisor Hahn had already instructed the county's chief engineer, Harvey Brandt, to investigate Olive View's ruins and determine whether the hospital had been built in compliance with local building codes. To do so, Brandt relied on assistance from California's Structural Engineers Association (SEA), a long-standing professional organization that had helped develop the state's earliest seismic safety codes. The SEA put together its own special task force of geologists, electricians, soil scientists, engineers, and architects to comb through the hospital's ruins.[66] By late May, barely three months after the deadly 1971 quake, the task force concluded that Olive View met local building standards but that those standards were too narrow and weak. In addition to undermining the hospital's foundations, the quake damaged many nonstructural elements of the hospital, such as elevators, fire alarms, and ventilators. It had dislodged air conditioning units, cooling towers, emergency generators, and even toppled a chemical tank.[67] It had also displaced lighting fixtures, power cables, and high-voltage conductors. Hospital safety depended on protecting these components of medical facilities. "If that fails, everything else fails," one engineer explained.[68]

On the heels of the SEA's recommendations, the county's Earthquake

Commission released its own highly anticipated and well-publicized report. The report began with a stern warning: "There is reasonable expectation that before the end of the century an earthquake of much greater magnitude will occur in Southern California."[69] To avoid an even worse calamity, the commission made fifteen proposals. Most importantly, it called for the reinforcement of large structures like dams and freeway overpasses as well as the renovation or demolition of all old buildings in the region. It also recommended new regulations for nonstructural elements of buildings, including all mechanical and electrical equipment, and the development of special building codes for critical structures like hospitals. "Certain types of structures and facilities which are particularly important in post-disaster operations such as hospitals should be designed and constructed to withstand strong earthquake shaking and yet be able to continue to function," the commission concluded.[70]

Local officials heeded the advice. Over the next year, the County of Los Angeles, under Supervisor Hahn's guidance, took notable steps to minimize the dangers of seismic activity. For starters, it adopted more stringent bracing requirements for new homes and buildings, and prohibited construction over faults. It also developed plans to finance the reinforcement or replacement of hazardous old buildings together with older bridges, dams, and other structures.[71] Additionally, the county introduced new building codes for facilities it considered vital to emergency response and relief operations, beginning with hospitals. In establishing these new regulations, county officials looked for advice from structural engineers and hospital administrators as well as representatives from the construction, banking, and insurance industries. They also consulted with officials from various state and federal agencies. Gradually, the parties began strengthening the framework for local regulation of earthquake hazards.[72]

The accomplishments in Los Angeles both paralleled and promoted reforms in the state capital of Sacramento, where seismic safety advocates had long pushed for disaster mitigation policies. In 1969, in the aftermath of the troubling quake along the San Jacinto Fault, California's Senate and House of Assembly had formed the Joint Committee on Seismic Safety. Chaired by Senator Alfred Alquist of San Jose, a Democrat and longtime supporter of public safety issues, the eight-member committee was well balanced and bipartisan. Once established, the committee divided itself into separate work-

ing groups, each of which drawing from its own pool of technical experts. While some groups focused on earthquake preparedness and land-use policies, others explored new developments in earthquake prediction research, structural engineering, and emergency response planning. The indefatigable Alquist coordinated and promoted the groups' work. His committee was in the midst of drafting new seismic safety legislation when the 1971 quake occurred.[73]

The quake presented Alquist with the perfect opportunity to launch the legislation. In its aftermath, the state senator introduced multiple bills designed to protect Californians from seismic hazards. One bill called for increased funding of earthquake research and engineering, and another proposed new restrictions on construction near fault zones. Others asked for more comprehensive disaster relief planning and for stronger seismic design requirements for structures deemed essential in times of crisis, such as hospitals. Still another proposed the establishment of a new state agency to oversee all matters related to seismic safety. Each of these bills met opposition from well-organized interest groups and came under intense scrutiny from both the fiscally conservative Senate Finance Committee and the California Department of Finance. Then, they also needed the signature of Republican Governor Reagan, a champion of small government. Reagan's cost-conscious administration had generally resisted funding any new government initiatives, regardless of their merits.[74]

The case of the Hospital Seismic Safety Act brings these challenges into sharper focus. When Alquist introduced legislation to strengthen seismic safety codes, he hoped to bring hospitals under provisions of the 1933 Field Act, which strengthened safety codes for schools. His plan called for the creation of a new agency—the Hospital Building Safety Board—to approve the planning and construction of all new hospitals in the state. The original proposal, however, threatened the authority of the California Hospital Association, a powerful coalition of health providers and their administrators. It also undermined powers of the Department of Public Health, the agency most responsible for the approval of new medical facilities. It worried cities and counties, too. California's surging population had left many communities in desperate need of new hospitals, and the proposed legislation promised to raise the cost of building those facilities. Despite strong support from scientists and engineers, the Senate Finance Committee killed the legislation in July 1971. The revisions Alquist made to the bill before reintroducing

it in 1972 testified to the influence of political coalitions in policy making. The revised legislation not only gave hospital administrators additional representation on the proposed safety board, but also placed the board under the direction of the Department of Public Health. The revisions did not satisfy everyone, but they proved sufficient to move the legislation through the senate and the assembly. They also pleased Governor Reagan, who signed the bill into law on November 11, 1972.[75]

The Hospital Act marked a major achievement in the history of seismic safety reforms. To be brief, the act required the governing boards of proposed hospitals to submit new development plans to an eleven-member board of highly trained professionals for approval. Those plans had to show careful consideration of geological conditions at proposed sites and demonstrate that both structural and nonstructural elements of hospitals could survive sudden and significant ground movement. The board had the responsibility of both ensuring that plans met safety standards and supervising the construction process. Though the act did not require existing hospitals to meet new standards, it stipulated that any alteration of those facilities had to comply with the standards. Legislators hoped that the normal replacement of older buildings would bring most hospitals up to code within twenty-five years.[76]

Implementing the Hospital Act proved more challenging than anticipated. The act did not say how strong seismic standards should be. The only guidance came from a clause in the law that said new hospitals had to be "designed and constructed to resist, insofar as practicable, the forces generated by earthquakes." The words raised a thorny question: Should the state set standards for hospitals irrespective of construction costs? Even seismic safety experts disagreed. Some largely dismissed the cost factor, insisting that hospitals be built to withstand the worst-case scenarios. Others thought it best to plan for "probable" quakes, which promised to hold down costs. The controversy remained unsettled until July 1974, when California's attorney general said the state's new hospital safety board would consider design and construction costs when setting the standards. The board then turned to the Structural Engineers Association for further advice, and the association recommended strengthening the standards for future hospitals by roughly 50 percent, depending on the characteristics of the soil beneath them. To put it all too simply, that meant further fortifying medical facilities against the energy generated by earthquakes.[77]

As these changes unfolded, Alquist introduced other seismic safety bills, including legislation that required cities and counties to identify and appraise hazards posed by earthquakes. These hazards had to be outlined in the "general plans" that communities drew up to address local problems, such as housing shortages. Governor Reagan not only supported the legislation but also formed his own council of experts to help implement it. The Governor's Earthquake Council soon presented local public administrators with a series of specific questions to answer. It asked about the location of all hazardous old structures in their communities, for example, and whether there were plans in place to demolish or rehabilitate those structures. It also asked about local land-use policies, zoning ordinances, and new housing projects. What's more, it asked administrators to explain how they planned to respond to a major earthquake: How would dead bodies be removed from the quake's ruins? How would debris be cleared from major thoroughfares? How would disaster victims be evacuated to safe places, and what did those places look like?[78]

Perhaps most importantly, Alquist pushed through legislation that established the California Seismic Safety Commission—a twelve-member board designed to keep the state abreast of new advances in seismic safety. Once again, Alquist made compromises to accomplish his goals. Simply put, he had to change the commission from a permanent, independent rule-setting agency to an advisory group whose life had to be renewed on a regular basis. Furthermore, he agreed to reduce the commission's initial funding by more than half. The commission nonetheless became an integral part of California's framework of codes, laws, and agencies that addressed the state's earthquake problem. The swearing in of its original members, on May 27, 1975, brought a sense of closure to a statewide seismic safety campaign.[79]

The federal government made its own contributions to this safety movement. Upon learning of the 1971 quake, the director of the Office of Emergency Preparedness (OEP), G. A. Lincoln, dispatched experts from the Corps of Engineers and US Geological Survey to investigate the structural integrity of all federal facilities in Southern California. As Lincoln told President Richard Nixon after the quake, seismic tremors would likely bring down other VA hospitals in the region. The only way to prevent such disasters, Lincoln emphasized, was to make those structures more earthquake resis-

tant.[80] Nixon, a native Californian, had a keen appreciation of the problem. The president had already declared the quake a major disaster, and as a result more than a dozen federal agencies provided services and resources to local relief and recovery efforts. He readily agreed that safety standards for VA hospitals should be strengthened.[81]

Congress put this plan in motion. Acting largely through the House Committee on Veterans Affairs, Congress sent several of its members to Los Angeles to consider the implications of the 1971 disaster. This subcommittee held widely advertised public hearings in the downtown federal building, where congressmen asked prominent seismologists and engineers whether they thought other federal hospitals in the area were safe. House Democrat Don Edwards of San Jose opened hearings by stressing the subcommittee's commitment to protecting all VA patients and employees. "We seek safeguards to minimize the danger inherent to VA structures," Edwards stated. "This we will do," he promised. "That is our mission."[82]

The mission proved daunting. As several experts told the subcommittee, seismic forces represented a threat to VA hospitals across the country. Bruce Bolt, the director of seismographic programs at the University of California, Berkeley, reminded Edwards's subcommittee that a few federal facilities in California stood dangerously close to major faults, as did facilities in Alaska, Montana, Nevada, Washington, and other states.[83] Caltech's George Housner warned that any hospital as old as the Sylmar facility would likely collapse if shaken by powerful ground movements. "[Sylmar] was in a condition such that it would be expected to collapse in the event of strong ground shaking," in Housner's opinion.[84] Henry Degenkolb, known for his own contributions to earthquake engineering, reminded the subcommittee that California was among the few states that had developed building codes with seismic safety in mind, and yet even its codes needed strengthening. "Much more needs to be done," the structural engineer insisted.[85]

In response to such warnings, the subcommittee recommended the reinforcement or replacement of all VA hospitals in earthquake-prone regions of the country. It also recommended new studies of all proposed sites for future VA facilities and careful inspection of construction practices on those sites. "Future proposed sites for VA hospital construction should be carefully examined by a competent team of experts," the subcommittee's report stated. "A rigid construction inspection system, preferably conducted by

independent experts, should approve VA hospital construction while it is in progress."[86] These were tall orders that required a dramatic expansion of the VA's responsibilities and budget.

Over the next year, however, the VA chipped away at the challenge. In April 1971 the agency established an advisory committee to investigate the vulnerability of its California facilities. Nine months later, just days before the one-year anniversary of the San Fernando disaster, VA Director Donald Johnson announced that numerous buildings at VA complexes in California had been deemed unsafe.[87] Investigators found thirteen unsafe buildings at a hospital east of San Francisco and another eighteen at a facility southeast of the San Francisco Bay. They found thirty dangerous structures at an old hospital in West Los Angeles, where many of Sylmar's displaced patients and employees had been transferred. As a result of these findings, Johnson ordered the transfer of nearly a thousand patients and many of the agency's employees to safer places. He also approved plans to demolish or modify the designated structures. As Johnson explained at the time, the VA intended to replace its Sylmar property with a new facility east of Los Angeles, in the growing city of Loma Linda. The agency would also build a new 880-bed facility in San Diego to replace its old West Los Angeles complex.[88]

In taking these steps, federal officials tried to assure the public that they had learned valuable lessons from the Sylmar tragedy. Director Johnson, for example, pledged to the public that the VA's new Loma Linda facility would have state-of-the-art safety technology. "Every known safeguard will be incorporated in the new 630-bed hospital approved for Loma Linda, and in all [other] VA hospitals to be built in the state."[89] By the time Johnson made those statements, the VA had drawn up plans to ensure all its hospitals would meet new earthquake-resistant design requirements appropriate for their geographic locations, and to keep its facilities operational after major disasters. It also established a permanent committee of seismic safety experts to offer advice on all such matters.[90]

The final act of this safety movement played out in 1977, when Congress passed the Earthquake Hazards Reduction Act. Introduced by Democratic Senator Alan Cranston of California, the act set up a new federal agency to investigate and reduce the risks of seismic activity. Cranston had pushed Congress to support seismic safety proposals since early 1972, when he introduced bills to fund earthquake prediction, fault mapping, and engineering research. Each of those bills had died in committees after congressmen

objected to their costs. To many if not most lawmakers the idea that earthquakes could be predicted seemed ludicrous, better left to soothsayers and psychics. Moreover, at a time when the nation was still mired in the Vietnam War, the problem of earthquake safety seemed relatively unimportant.[91]

Senator Cranston and his advisors persisted, and the political landscape soon changed in their favor. In 1976 a series of devastating earthquakes in China, Italy, and Guatemala reminded lawmakers of how costly seismic activity could be. Moreover, earthquake specialists had recently discovered a "bulge" along the San Andreas Fault, near the town of Palmdale just north of Los Angeles, which they suggested was a sign of an impending disaster. A similar phenomenon, such specialists warned, had preceded other large quakes. These developments moved political needles. President Gerald Ford's domestic advisors, including Vice President Nelson Rockefeller, set up a new panel of experts to study recently proposed earthquake mitigation plans. Led by Nathan Newmark, a distinguished engineer from the University of Illinois, the panel cleared the way for passage of Cranston's hazards reduction act, which established the still-standing National Earthquake Hazards Reduction Program to coordinate local, state, and federal efforts to reduce losses caused by quakes. The legislation had bipartisan support. President Jimmy Carter signed it on October 7, 1977.[92]

The benefits and limitations of this seismic safety movement remained unclear for more than two decades, until another powerful early-morning earthquake rocked the San Fernando Valley. On January 17, 1994, a quake centered beneath the neighborhoods of Northridge and Reseda killed fifty-seven people and injured some sixteen hundred others seriously enough to be hospitalized. Frighteningly reminiscent of its 1971 predecessor, the quake wrecked homes and businesses, rendered freeways and city streets impassable, and knocked out electricity, gas, water, and telephone services. The heavy shaking also sparked fires, explosions, and even landslides. It damaged several hospitals, too, including the large VA facility in Sepulveda and two of the valley's newest medical treatment centers. Overall economic losses resulting from the quake reached approximately fifty billion dollars, making it one of the costliest disasters in American history.[93]

Despite the damage, the 1994 earthquake revealed the effectiveness of seismic safety regulations. For nearly ten minutes, the quake and its aftershocks shook one of the nation's most densely populated areas. At the time

it hit, some 9 million people lived in greater Los Angeles, or about 2 million more than were residing there in 1971. Roughly a quarter of those people had settled in the San Fernando Valley, pushing the valley's population from 1 to 1.5 million. Thousands of new homes had been constructed in the area during those years, along with scores of office buildings, shopping malls, and other large structures. Yet structural failures accounted for only thirty-three of the fifty-seven deaths, and nearly half of those deaths were attributable to the collapse of a single low-rise apartment building.[94] Post-quake investigations concluded that the reforms of the 1970s had not only saved lives but also minimized damage to electrical power plants, water treatment systems, bridges, and other vital structures.[95] More telling was the limited damage to local hospitals. Though the quake ruptured many nonstructural elements of those places, such as water lines and temperature control systems, even the hardest-hit facilities were fully operational within a few weeks, including the VA's hospital in Sepulveda. The quake did not kill, injure, or displace any hospital workers. Through any lens, that looked like progress.[96]

These reforms were well overdue. By the time they took shape, earthquakes in California had repeatedly underscored the limitations of the state's seismic safety codes. Support for seismic safety regulations increased notably in the late 1960s, but it took the 1971 quake to bring national attention to California's seismic safety problem. The quake was unusually damaging and frightening. It stirred up everyone's fears and insecurities, and it raised the most troubling questions: Is another, larger quake coming? If so, how many more homes and places of employment might collapse? How many more people might perish? Prodded on by an anxious citizenry, a few dedicated public officials responded to such questions with the seriousness they deserved. Advocating for everyone who bore the risks of earthquakes, policymakers found practical solutions to pressing problems. In doing so, however, they relied on the advice of prominent scientists, engineers, and other highly trained professionals who understood seismic safety issues. The media played a crucial role in this movement, too. News coverage highlighted the damage that quakes can cause and kept the public informed of new seismic safety proposals. Though the movement had special consequences for California's hospital workers and their patients, it ultimately benefited the general public. Yet such victories have always proved limited or fleeting. Seismic safety remains a source of high anxiety in California and other quake-prone parts of the country.[97]

5

MGM Grand Burns in Las Vegas!

This miraculous and magical world is also demonic and terrifying,
swinging wildly out of control, menacing and destroying blindly as
it moves.

Marshall Berman, *All That Is Solid Melts into Air*

On November 21, 1980, at 7:15 a.m., fire broke out in a delicatessen
on the ground floor of the crowded MGM Grand Hotel in Las Vegas, Nevada,
a posh twenty-six-floor resort on the world-famous Las Vegas Strip. The
blaze quickly spread to the adjacent 140-yard-long casino and then to ground-
floor shops, offices, and reception areas. Just as quickly, thick black smoke
climbed throughout the resort's three hotel towers through the air conditioning system and seeped into the rooms of tourists, most of whom were
still asleep. "When we woke up, we couldn't see a thing," one couple recalled.
The conflagration claimed the lives of eighty-five people, making it one of
the deadliest hotel fires in American history.[1]

Over the next few days, the gruesome details of the tragedy came to
light. News reports told of bodies piled atop one another in locked stairwells
and elevators. They described people trying to scale down the hotel's walls
using makeshift ropes, only to fall to the ground alongside the heavy pieces
of furniture that other desperate guests used to break sealed windows and
for fresh air to breathe. Published photographs made the tragedy all the
more graphic with images of smoke and flames billowing from the hotel and
terrified tourists huddled on balconies or dangling perilously from the cables of rescue helicopters. Distraught people were seen weeping over dead or
damaged bodies of friends or relatives. "An anguished hotel guest grieves

over his wife," read the caption below one newspaper photograph.[2] News coverage of the fire included compelling stories of survival as well as tragedy. Reporters interviewed hotel guests who had scrambled down stairwells fearing for their lives or climbed their way to the rooftop hoping to be rescued. The media also speculated on the actual or potential impact of the fire on the MGM's future and on Las Vegas tourism in general.

These were certainly essential aspects of the episode, its aftermath, and definitely deserved attention. Another major dimension of the story, however, was relatively neglected by the news media: the impact of the fire on MGM employees. Until the fire damaged and closed it for eight months, the MGM was Nevada's largest private employer. More than forty-five hundred people worked there, about a third of whom were on duty when the fire started. Though only nine of those employees lost their lives in the fire— two maids, two room-service waiters, two security guards, a baker, a cashier, and a slot machine mechanic—smoke inhalation left many others hospitalized. Still more employees were severely shaken by what they saw and experienced during the fire and its immediate aftermath. "I can't put it out of my mind," one of the latter group said.[3] "It was horrible," another recalled. "I saw people jump. I saw them lying dead on the street."[4] Some needed psychological counseling to deal with the experience. "I went to a doctor the following week," one person said. "He not only gave me sleeping pills but anti-depressants."[5] Like other workers displaced by disasters, the employees struggled mightily to cope with a host of problems and challenges after the fire. The fire not only wiped out jobs and incomes but also disrupted daily routines and strained family relationships.[6]

The story of the MGM fire offers yet another chance to meditate on the unintended, solemn consequences of material developments and to show how disasters have hastened the rise of modern safety regulations. Picture yourself relaxing high above Las Vegas in a luxurious hotel room, only to get trapped in a rapidly spreading fire. Would you have remained calm and collected, logical and levelheaded? What if people started throwing furniture through windows and screaming for help? This was an unusually terrifying and devastating fire. In its aftermath, Nevada passed the nation's most stringent laws concerning fire safety in high-rise buildings. But these laws did not just sail through legislative chambers in Carson City, the state's capital. Complying with safety proposals meant retrofitting the state's large resorts with new sprinkler systems and other firefighting devices, and thus prom-

Fire breaks out at the MGM Grand Hotel in Las Vegas, November 20, 1980. The fire killed eighty-five people, including nine of the resort's employees. It also displaced the resort's giant workforce for several months. Getty Images

ised to be costly and disruptive. Employers generally resisted such proposals, often on the grounds that the government should stay out of their affairs. Those who supported the new measures usually did so grudgingly, without believing workplace safety was a vital part of their long-term business plans. In Nevada tourism, however, resort owners realized that public safety concerns threatened their long-term interests. Working through the Nevada Resort Association, one of the state's most powerful lobbying groups, the owners came to support legislation that required costly new devices, so long as the legislation applied to all resorts equally.

The MGM fire took place against the background of a newly emerged metropolitan center and the still expanding tourist industry. Historians have had much to say about the importance of Las Vegas and tourism in the postwar

Southwest and have shown how the city and industry became one of the region's leading engines of economic and urban growth. In the 1950s and '60s, when gas was cheap and new interstate highways crisscrossed the region, millions of vacationing families traveled west past places like Carlsbad Caverns and the Painted Desert before ultimately reaching Disneyland, the "happiest place on earth." Along the way, many stopped in Las Vegas to gamble and relax in its casinos and resorts. Growing numbers of Californians also traveled to Sin City, as Las Vegas was known, either by car or plane, simply to enjoy a weekend or holiday. Las Vegas became the Coney Island of the West during these years, a place to partake in forms of recreation and self-indulgence that diverted its visitors from the routines of everyday life. Its glitz and glamour stood in sharp contrast to ordinary workplaces and neighborhoods. Where else could one see representations of pagan statues, ancient pyramids, and medieval castles, or even gigantic, flashing neon signs? Where else could they openly gamble?

The rise of Las Vegas was an unusually striking development in this postwar period. In the early 1940s, Las Vegas was still a whistle-stop on a railroad to Los Angeles. Many of the people who lived there had helped build the nearby Hoover Dam on the Colorado River and then found jobs in the local building trades and service industries. Many others had moved to the greater metropolitan area to work at a large, federally funded wartime plant ten miles southeast of Las Vegas, where they processed magnesium ores— natural mineral formations used in making military aircraft and incendiary bombs. Southern Nevada's population surged in the early postwar years, when better distribution systems and the advent of air conditioning made the desert climate more hospitable. Still, at midcentury, only fifty thousand people lived in greater Las Vegas. By comparison, the population of greater Los Angeles topped four million.[7]

The wheels of commerce in Las Vegas turned quickly in the 1950s, when entrepreneurs and their financial backers opened several multimillion-dollar resorts along a stretch of highway just beyond the city's southwestern limits, which soon became known as the Strip. Casinos like the Sahara, the Sands, and the Stardust were much larger and more elegant than their older downtown counterparts and offered more than just a chance to relax and gamble legally. They provided their guests with an array of special services and amenities, including championship golf, elaborate stage shows, and world-renowned entertainers. The rise of these luxury resorts encouraged

owners of downtown properties to remake their properties. By the end of the decade, older landmarks like the Golden Nugget offered tourists many of the same amenities as Strip properties, but they never matched the panache of casinos on the Strip. The downtown area nonetheless made an essential contribution to this emerging tourist economy.[8]

The pace of development remained heady in subsequent years, especially after Nevada amended its gaming codes to attract the capital resources of large corporations. The arrival of Howard Hughes in the city in 1966 marked the symbolic beginning of the "corporate revolution," when control of gaming passed from family-owned enterprises to less personal, hierarchical business structures. Over the next four years, Hughes purchased six Strip properties that collectively accounted for a quarter of the Strip's revenue. In the early 1970s, after Hughes left the city and set up the Summa Corporation to oversee his Nevada investments, large hotel chains like the Hilton and Hyatt Corporations made their own investments in Las Vegas and were pouring resources into them. The Dunes, the Sands, the Flamingo, and other iconic resorts launched major construction projects in these years.[9] Meanwhile, a number of smaller resorts sprang up on or around the Strip, including Bourbon Street, the Barbary Coast, the Imperial Palace, the Maxim, and the Marina Hotel.[10]

By 1975 the revolution had largely run its course. Las Vegas was now a major Sunbelt community and a national tourist destination. More than four hundred thousand people lived in the metropolitan area, whose annual tourist count topped nine million. Roughly half of the area's visitors now arrived by air, landing at the ever expanding McCarran Airport. Compared to urban communities in the East and Midwest, where industrial development had stalled, the local economy seemed phenomenal. New banks, warehouses, and shopping malls were everywhere. New homes were priced relatively low and kept the construction industry strong. But gaming and tourism remained at the heart of economic life. More than fifty thousand people, or nearly a third of the local workforce, earned a living in tourist-related businesses. About a quarter of these people worked in large hotels and casinos, mostly those along the Strip. Properties like the Desert Inn and the Sahara each employed about twelve hundred people, and Caesars Palace had more than two thousand. In the employment hierarchy, a small circle of corporate officers held control of business operations.[11]

Resort workers fell into four general and often overlapping categories.

One category encompassed professional and semiprofessional employees who held jobs that required administrative and technical skills, and even creativity. This group included not only security managers and chief engineers but also executive chefs, bandleaders, choreographers, and other supervisors within the workforce who earned salaries commensurate with their training and experience. A second and larger category incorporated white-collar groups such as secretaries, desk clerks, and switchboard operators. Their jobs typically demanded technical training as well as dexterity, good judgment, and the ability to work with other employees and even resort patrons. A third, larger group consisted of blue-collar workers. Some of these employees worked outdoors, such as gardeners and parking lot attendants. Others, including slot machine mechanics, operating engineers, and security guards, remained indoors. Though some blue-collar jobs required relatively few skills, they all demanded some level of expertise and initiative. The largest occupational category included the employees who dealt directly with customers, such as bartenders, cocktail waitresses, dealers, baggage handlers, and housekeepers. Like many of their blue-collar counterparts, these employees were not highly skilled, but they had to deliver services dependably, correctly, and courteously.[12]

The mammoth MGM Grand, the newest and most imposing resort, reflected these historical forces and patterns. Located at the corner of the Strip and Flamingo Road, the MGM employed about forty-five hundred people, or more than twice as many as any other local resort. The extraordinary, all-inclusive property was part of the business empire of Kirk Kerkorian, who had made his fortune in the airline industry. Kerkorian had financed the resort's construction by selling his interests in the Flamingo and the new International Hotel to the Hilton Corporation. When the MGM opened in 1973, it advertised itself as the world's largest resort, with 2.5 million square feet of covered space, 2,076 hotel rooms, and many stunning amenities. Its casino encompassed an area the size of a football field and included more than a thousand slot machines and a hundred gaming tables. The resort's seven kitchens could serve thirty thousand meals a day; its two large showrooms featured star entertainers and lavish stage productions.[13]

The physical setting was magnificent but not fireproof. It had few of what are now standard fire safety features in high-rise buildings, including automatic sprinklers and smoke detectors. Its ground floor, where the fire started and spread uncontrollably, was filled with furniture and ornamental objects

that were especially combustible, and the firefighting equipment in the vicinity was hard to find. The chairs and booths in the deli, where the fire initially came into view, were covered with synthetic materials that emitted large amounts of heat and smoke when ignited. So, too, were the casino's walls and ceiling ornaments, including the room's large "crystal" chandeliers. The casino's low ceilings and absence of windows were also problematic. Bounded by the windowless casino, the fire quickly consumed the room's oxygen, and thus the process of combustion nearly ceased. As oxygen reentered the casino through the resort's main entrance, however, it ignited the fire's unburned fuel and gases, causing a massive explosion. To make matters worse, the resort's elevators, laundry chutes, and other air ducts were designed in a way that facilitated the spreading of smoke through the three towers of the T-shaped hotel. Though the towers had six stairwells, only three of them functioned as emergency exits. The doors of the other three could be opened from the inside on the top and bottom floors only. This security feature prevented some guests who fled their smoke-filled rooms from escaping to safety.[14]

The MGM also had no emergency evacuation plan. The only directives for evacuating the hotel were posted on a bulletin board in the housekeeping department and thus beyond public view. In any case, the plans were addressed to employees, not tourists. They directed employees to assemble in a rear parking lot, where supervisors would ascertain that all employees were safe. They also mandated that employees follow "house rules" designed more to prevent fires than to prescribe conduct in case of an actual fire. The rules were trite to the point of meaninglessness: "Take every precaution to prevent fire," one of them read. "Throw all cigarette butts, matches, oily rags, and any other rubbish in properly provided receptacles." "In case of fire," another explained, "the first consideration is the safety of all guests and employees; then attention can be given to fighting the fire." Still another house rule instructed employees to protect resort property during a fire: "Place loose papers and records in the file cabinets. Close all drawers and cabinets." The rules even carried a punitive tone: "Any infraction of these rules may result in disciplinary action."[15]

Two days after the fire, the *Las Vegas Sun* revealed that the owners of the MGM had ignored the suggestions of local fire department officials when building the giant resort. During the early stages of the construction process, Clark County Fire Marshall Carl Lowe advised the owners to install a

comprehensive sprinkler system. According to Lowe, the owners claimed the system was too expensive to install and chose instead to meet Nevada's minimum requirements for fire safety. They installed sprinklers only in the resort's basement, large showrooms, and a special room for high rollers on the top floor. "All they wanted to know was what the code stated," Lowe said. "A fireman can see a lot more than the average citizen. But with the builders," Lowe added, "all they saw were dollar signs."[16] Lowe suggested that the state's building codes were not only outdated but also not rigorously enforced. Resort owners had pressured county officials to overlook code violations, he suggested, all in the name of profits. "They don't want to put out the money," he told the press. The *Sun's* report fueled speculation that the MGM was solely to blame for the fire and that other large resorts in Nevada might also go up in flames.[17]

Investigators initially suspected that kitchen workers in the delicatessen had started the MGM fire by allowing grease to collect in a ventilator but later traced the fire's origins to faulty wiring in the delicatessen's walls. An improperly installed refrigerator compressor apparently overheated combustible material and thereby created an electrical malfunction. Wiring simmered for hours undetected and eventually set fire to the plywood walls. By the time anyone noticed the fire, smoke and heat had spread through an open attic above the deli and the casino. When the fire broke through the deli's ceiling, oxygen rushed through the entire attic area, turning already glowing cinders into flames that quickly consumed the casino. "It was a classic case of a fire that built and burned," Clark County Fire Chief Roy Parrish later explained.[18]

Ironically, two firefighters on vacation from Illinois were having breakfast in the coffee shop next to the delicatessen when the fire began, but they were unable to offer much help. As the men later told investigators, a security guard approached them at their table saying the room had to be evacuated as a precautionary measure. When the firefighters saw smoke and flames in the delicatessen, they immediately offered their assistance. "We identified ourselves as firefighters and asked for a fire extinguisher or hose line and [said] we'd help knock down the fire till the fire department got there," one of them later told investigators. Surprisingly, the guard did not know where to find an extinguisher or hose. "He said they were not accessible," the fireman explained. The two men then went looking for firefighting

devices themselves. "We walked all the way back to the north wall and then back to the east past the elevators looking for a fire extinguisher or hose, and we couldn't find one." By the time they got back to the deli, a low-hanging cloud of smoke had engulfed the entire area. Understanding the danger they were in, the firefighters instructed people fleeing the scene to lower to the ground and head for the fire exits before doing so themselves.[19]

Tim Connor, an engineer conducting his morning inspections, was one of the few employees who tried fighting the fire. As Connor approached the coffee shop where the two firemen had been, he saw what he later described as a "sheet of flames running from the top of the [deli's] counter to the ceiling." The engineer reacted quickly, first using the nearby intercom system to contact security. "I went directly to the telephone, which is back at the entrance behind the desk," he later explained. "I told Security they had a fire in the Deli and he asked me if it is bad enough for the Fire Department, and I said, 'Hell yes, roll em!'" Unlike security guards, Connor knew that a fire hose cabinet was outside the Barrymore, the fine-dining restaurant next to the delicatessen, in a recessed area of the wall. By the time he reached the spot, however, the fire had spread. As he broke the cabinet's glass face, a blast of hot air knocked Connor to the ground. "I was totally engulfed in smoke," he recalled. "I could see the floor, but I couldn't breathe." Fortunately, the employee knew he was near a stairwell. "Knowing the hotel as I do, I went up the stairs, across a little parapet and hit the doors there, which were fortunately unlocked."[20]

The experience of Harvey Ginsburg, a local resident who fled the fire, was illustrative of the larger pattern. Ginsburg was walking through the casino on his way to the coffee shop that morning when he smelled smoke. "I noticed a foggy, smoky atmosphere, especially in the ceiling area of the casino," he told investigators. "My first impression was the air conditioning system was not working properly. At this point I did not anticipate that the smoke was a result of fire, since business was proceeding as usual in the casino." Continuing toward the coffee shop, however, Ginsburg saw flames shooting out of the delicatessen and people running toward him: "It appeared as though panic was setting in, so I turned around and trotted approximately halfway back out of the casino." For a brief moment Ginsburg felt safe, but when he turned around, he saw fire racing toward him: "The flames had already entered the casino area and appeared to be a swiftly rolling wall of flame reaching from floor to ceiling." Ginsburg ran as fast as he

could through the casino, along with a wave of gamblers and dealers. "While I was running other people started to run," he recalled. "Chairs from the blackjack table were strewn in the aisle. I tripped on one of them but still managed to get out of the casino in about 20 to 25 seconds."[21]

Gamblers and dealers in the casino recalled similar scenes. When the gamblers initially smelled smoke, many were hesitant to leave their slot machines and table games, especially those who were winning. One who had come to Las Vegas all the way from Rhode Island later admitted that he tried to exchange his chips for cash before coming to his senses. "I went over to the cage to cash out and I couldn't breathe," the man remembered. Realizing his foolishness, the man and his friend quickly raced toward the front entrance. "We ran out the casino and looked back and the canopy was on fire. The whole joint was an inferno."[22] Other gamblers wasted no time abandoning the place. "I left my chips and everything," one of them explained. "It wasn't worth it to try and keep them. People were running out of the room and panicking. Everyone came out crying and said to get out of there."[23] As more smoke entered the casino, pit bosses ordered dealers to "put a lid" on their games, which meant covering house chips with large glass panels stored under their tables. The dealers then ran for the entrance doors.[24]

Meanwhile, administrative personnel directly above the casino were also fleeing the resort. One of those employees, Louis Miranti, later told fire investigators about his own harrowing flight to safety. Miranti was part of the security force that worked directly above the casino floor, in the "eye in the sky," where management scrutinized the habits and routines of both gamblers and casino workers. He had just returned to his third-floor office to write a daily report when he heard slight "crackling noises." Miranti stepped out of his office and saw what he later called "a billow of smoke, like a tidal wave coming towards me." He reacted quickly. "I ran to the hallway leading back to the executive offices and the executive elevators." Finding the elevator doors closed, Miranti descended a nearby stairwell before realizing the fire was raging below him. "I felt the heat," he told investigators. Miranti then ran back up the stairs into the now smoke-filled hallway. Using his shoulder as a battering ram, Miranti broke through a set of double glass doors to get to a stairwell in the auditing department, located at the southwest corner of the resort. With smoke at his back and quickly losing consciousness, Miranti finally reached the stairs and stumbled down them. "I had my hand out in front of me trying to protect myself while I was running, so I wouldn't hit

anything," he recalled. When he finally reached ground level, a security guard was there to help him. "I felt someone grab my hand and pull me out."[25]

On the resort's upper floors, where the fire claimed most of its victims, hotel guests were waking up to find that their worst nightmare was a reality. The elevator shafts had turned into smokestacks, bringing torrents of blinding, choking smoke into hallways. Most people tried to survive the fire by remaining in their rooms and shoving towels beneath doors. "We put mattresses against the wall and kept stuffing towels under the door to keep the smoke out," one of those survivors explained. Others pressed something over their faces and rushed to the closest stairwell. "People were running through the halls without any clothes on," one survivor remembered. "They were just wearing towels and bed sheets they'd grabbed." Even those that made it into the stairwells were still in danger, for some of those escape routes led directly to the most dangerous areas of the resort.[26]

The experiences of Marv and Carol Schmatzman, who had rented a room on the resort's sixteenth floor, were especially chilling. Unlike most tourists, the Schmatzmans were already awake when the fire began, packing their suitcases for an early-morning return flight to St. Louis. When the couple saw smoke seeping into their room, they grabbed their belongings and fled. They found a stairwell in the smoke-filled hallway and ran down to the ninth floor before encountering a frightened crowd. "More people kept getting in and there was no air, and you couldn't hardly see, and you could hardly breathe," Marv recalled. When someone said the fire was on the ground floor, the couple turned around and headed toward the roof, as a few other people had already done. The climb to the top proved extremely difficult and required stepping over several exhausted tourists. "A lot of people were sitting down because they had given up, and we had to crawl over them," Carol remembered. By the twenty-fourth floor, Carol was exhausted, losing all hope. "When you inhale smoke it makes your extremities very weak," she explained. Pulling themselves along the handrail, the Schmatzmans finally reached the roof, where they and others waited anxiously to be rescued.[27]

Local firefighters arrived on the scene within minutes after the fire started. One of the first to arrive, Bert Sweeny, later described how he and his crew tackled the fire. When the crew arrived, the men saw little evidence of impending disaster. "When we pulled in, there was no smoke visible, there was

Statues smolder in front of the MGM Grand Hotel, Las Vegas, November 20, 1980. A ball of fire raced through resort's big casino and blew out the main entrance. Getty Images

no rush of people trying to get out of the hotel," Sweeny reported. "We put on our high-rise gear, walked in through the swinging doors, and went down to the exit-entry ramp to the casino floor." There, Sweeny saw people rushing toward him, fleeing the thick black layer of smoke in the casino. "We noticed this stratified layer which was probably down about 6 to 8 feet from the ceiling," he recalled. When the "fringes of flame" turned into a "fireball,"

Sweeny's crew retreated from the casino and set up a "line of attack" beyond its main entrance. As they did so, oxygen that had reentered the casino ignited unburned gases, blowing the resort's entrance doors open and engulfing the canopy that covered the area.[28]

With their fire hoses open, the firefighters rushed back into the casino and fought the blaze for five minutes, until their hoses ran dry. By then, other firefighters had tapped hydrants outside the resort and were fighting flames in the casino and the hotel's registration area. Several ran up stairwells into the building's three towers, where they encountered smoke so hot it blistered their face shields. Along the way, they had to step over the bodies of people asphyxiated by the smoke. By the time they reached the MGM's upper floors, fire trucks from across the metropolitan area had converged on the resort. Despite this prompt response, it took nearly two hours to bring the fire under control. During that time more than eighty people had died, mostly from smoke inhalation and related exigencies. Another seven hundred had been injured.[29]

The death toll would have been higher if not for the efforts of other rescue workers. The air force immediately dispatched nine helicopters from nearby Nellis Air Force Base to airlift people from the resort's roof, where hundreds had fled to escape the smoke and flames. In what they later described as a "dicey operation," pilots maneuvered the helicopters through clouds of black smoke and dropped "sling-extracting devices" to lift people into the aircraft. Seven or eight individuals at a time made the brief flight to a landing area adjacent to the resort. More than a thousand people made the trip safely, including at least twenty from hotel room balconies.[30]

Ambulance drivers rushed more than five hundred people to local hospitals, which were soon overwhelmed by the crisis. Sunrise Hospital, the region's largest medical facility, called in off-duty nurses and other medical personnel to treat incoming casualties, relying largely on volunteers to get patients from ambulances to emergency rooms. Some medical facilities had to turn away patients. "Every bed is full," a supervisor from Valley Hospital told a reporter during the day. "We even had to request four respiratory therapists from our Panorama City Community Hospital in California." Unless patients had urgent problems—broken bones, serious burns, or heart attacks—most of them were directed to the Las Vegas Convention Center, where paramedics and other medical workers hastily organized a relief station.[31]

Other agencies and institutions offered whatever assistance they could.

Firefighters inspect the MGM Grand Hotel's scorched casino, November 20, 1980. The windowless casino's low ceiling and flammable ornaments fed the fire. Associated Press

The Clark County School District, for example, provided fifty school buses to transport victims to aid stations and hospitals and to get unharmed survivors to McCarran Airport. The University of Nevada, Las Vegas, provided language instructors to help foreign tourists, and it offered shelter to victims in one of its gymnasiums.[32] The Red Cross and Salvation Army helped in the emergency aid center, contributing bottled water, blankets, and clothing. The Las Vegas Police Department facilitated the flow of traffic in the area, which was clogged with rescue vehicles and stunned onlookers.[33]

Local ironworkers also played a critical if fortuitous role in the crisis. A crew of about seventy ironworkers had been assembled outside the MGM at the time of the fire to begin work on a construction project. Some of the workers helped firefighters unravel hoses and carry firefighting equipment

into the resort; others accompanied firemen up smoke-filled stairwells to rescue stranded guests. "My arms and legs were hurting like hell from carrying people, but I couldn't stop," one of the latter group recalled. "I carried people out until my arms were too sore to carry any more," another said. A few of the men saved lives by riding electrically powered "spider cages" up and down steel cables hung along the exterior of the MGM's east wing, plucking people from hotel balconies. The scaffolding on which these cages rested could have collapsed under the weight put upon it. "I went up one time and was told not to go again," one of the men said. "But when I looked up at the building and saw the faces of those people, scared for their lives, you know someone has to do something."[34]

An untold number of MGM employees also risked their lives to save others. Housekeeper Daniel Sullivan, for example, rousted dozens of sleeping guests before being forced to leave the hotel for his own safety. As one guest remembered of Sullivan, "He ran up and down the corridor banging on doors and yelling 'Fire! Get out! Get out!'" "He wouldn't leave until everyone got out," the guest added.[35] Luke Carr, a lifeguard, helped ambulance drivers and paramedics turn the MGM's swimming pool area into a makeshift hospital, and he then assisted them in treating victims of smoke inhalation. Once firefighters had the blaze under control, Carr directed rescue units to the upper floors of the resort, where they went from room to room looking for survivors.[36]

Other employees acted according to the fire safety rules provided, protecting their employer's interests in a number of ways. Two pit bosses in the casino, for example, scooped up money in the cashier's cage and placed it in a fireproof vault before fleeing to safety. Employees in the "counting room" hastily put stacks of paper money, bookkeeping records, and other valuables in vaults before leaving the casino. A cashier in the coffee shop also pulled a tray from her cash register containing more than two hundred dollars and carried it with her when she left her workplace. As the smoke cleared, security guards kept possible looters from the casino and from abandoned rooms as firefighters extinguished the blaze. They also helped guests retrieve personal possessions they had abandoned in fleeing the burning hotel. Though many guests reported losses of jewelry and cash as a result of the fire, police arrested only one person for looting. The MGM itself lost only a small amount of cash, most of which had burned in the keno lounge.[37]

↓

Rays of light shine into the charred MGM Grand Hotel, November 20, 1980.
Associated Press

The fire forced the MGM to close for the foreseeable future and thus to lay off the vast majority of its employees. It destroyed not only the resort's casino but also most of its dining areas, lounges, and shops as well as its administrative offices and registration area. Although the fire caused little damage to the resort's two theaters, it ruined an assortment of costumes, props, and other supplies critical to showroom productions.

The damage was extensive and may at first sight have appeared even more catastrophic than it actually was. Republican Governor Robert List, who walked through the scene of the fire the day after it occurred, compared what he saw to the "bowels of hell." The fire, he told reporters, had reduced the resort to "unbelievably twisted contortions of metal."[38] The MGM's three top executives—Alvin Benedict, Fred Benninger, and Bernard Rothkoph—were fishing off the coast of California at the time of the fire, so management's response to the crisis seemed tardy and ineffective if not cold and uncaring. The executives finally held a press conference shortly after Gover-

nor List visited the property. In a prepared statement, Chief Executive Officer Benninger repeated the governor's assessment of the damage and expressed sympathy for disaster victims. "Words are inadequate to express our anguish," he told reporters. Benninger also said the MGM intended to rebuild and reopen the resort as soon as possible, perhaps by July 1981. He assured the press that the renovated resort would be safer than ever. He refused to say, however, whether its hotel rooms would be equipped with automatic sprinklers and fire alarms. Dodging that question, Benninger simply said, "The old rooms will have to be in compliance with building codes in existence today."[39]

As the press reported, this news conference was emotionally charged. Reporters shouted increasingly hostile questions to Benninger and his management team. Why had the MGM not installed better firefighting equipment? Why had it not even considered the matter? Repeatedly, Benninger reminded the press that the resort met all fire safety codes at the time it was constructed. In 1970, as he pointed out, state codes required neither smoke alarms nor sprinkler systems. Recent revisions to the code, he also explained, applied only to new buildings, not the MGM. When pressed on why fire alarms in the resort had reportedly failed, Benninger created more tension in the room by suggesting the failure "may have been a blessing in disguise." If the alarms had worked properly, he implied, the death toll would have been higher: "A lot of guests would have gone into the hallways and suffocated from the smoke. Staying in the rooms may have saved them." Benninger's words added to the sense of shock and outrage in Las Vegas and underscored the need for policy change. They also highlighted the role of the media in this story. By keeping the problem of fire safety before the general public, the media played a vital role in generating support for safety reforms.[40]

The closure took place in a national environment of economic contraction characterized at the time by such terms as downsizing, outsourcing, and other management innovations that were generally detrimental to labor. Despite the national economic situation, industry leaders in Las Vegas believed that tourism was largely immune to these economic trends. By the time of the MGM fire, however, that belief was called into question. The local tourist count had stalled in 1980, and the growth of gaming revenue in Nevada failed to keep up with inflation. As a result, several resorts postponed construction projects while others discounted room prices, announced hiring

freezes, and discharged expendable employees, including dozens of non-union dealers and other casino workers. Smaller restaurants and motels also discharged workers, further constricting employment.[41] Unemployment rose noticeably, to about the national average of 7 percent, and it jumped to nearly 9 percent in the aftermath of the fire. At the same time, rising inflation exacerbated the problems of workers idled by the fire. Between 1976 and 1980 the consumer price index rose nationally from a discomforting 6 percent to a punishing 13 percent annually, while oil prices increased dramatically.[42]

It was under such uncertain circumstances that the laid-off MGM employees pondered their future. "What am I going to do?" one of them asked. As the bewilderment they expressed suggests, most employees were unprepared for any long period of unemployment. They had bills to pay and families to feed and few resources to fall back upon. "My wife's pregnant and we're hurting," one waiter explained. "I'm frightened," another displaced employee admitted.[43] Within the first few days after the fire, workers learned that the MGM would keep about five hundred security guards and a few dozen other workers—chosen on the basis of seniority—on the payroll to help clean up rubble. All others collected the pay due them and sought whatever options they could find.[44]

Most of them filed claims for unemployment benefits. For the purposes of processing those claims, the Nevada Employment Security Department divided the laid-off workers into three groups, receiving and assessing the claims of each group over a three-day period. Despite this effort to bring order to the process, the claims office was crowded and chaotic for days. The experience was troublesome enough to inspire several newspaper editorials to deplore its delays and inconveniences. "The line at the employment office extended one-and-a-half blocks from the entrance," a frustrated applicant complained to the press. "The wait outside in the cold wind was two to three hours." The situation inside was worse. "It was a madhouse in there," someone said. "One guy punched another out for stepping in front of him in the line." Some claimants were so exasperated by the situation that they delayed their applications until the rush was over. "I'll just wait until next week when things are more quiet and fill out the papers," one of them said.[45]

The benefits that eventuated from this process were helpful but hardly adequate for anyone, especially a sole breadwinner with a family to support over an extended period. Like other states, Nevada based unemployment

benefit levels on percentages of lost wages, not including tips. For the typical worker displaced by the MGM fire, this meant a weekly check of only one hundred dollars (worth about three hundred dollars in 2020), or less than half of what such a worker earned previously. Many supplemented this income with food stamps, but workers unaccustomed to the rituals of welfare dependence often found the application process difficult if not demeaning, and at times it was intolerable. At least one woman found it degrading enough to abandon it. "I waited in this line for food stamps, and when I got up there they humiliated me," she recalled. "I told them I would never come back there, even if I starve to death."[46] Such words revealed the limitations of the modern welfare system along with the desperation of many laid-off workers. After major disasters like the MGM fire, the resources of welfare agencies often seemed inadequate, and welfare workers were sometimes viewed as impatient, insensitive, and even rude.

Two-thirds of the displaced employees belonged to trade unions, and they looked to their unions for help. The unions, however, had no means of dealing with the myriad problems created by the MGM fire. They existed to deal with employer–employee relations, not an industrial catastrophe. As such, the unions had modest strike funds and health and welfare programs, but the latter was only equipped to address individual needs, not systemic failures. The unions did work to facilitate the relief efforts of volunteer organizations like the Red Cross and the Salvation Army, and they helped state and federal agencies with emergency assistance. Their primary contribution to the plight of workers displaced by the fire was to assist them in finding new jobs. That was a daunting task. Lines at union hiring halls had been long even before the fire, especially at the Culinary (Local 226 of the Hotel Employees and Restaurant Employees International Union), to which most of the displaced workers belonged. The nearly four thousand people who suddenly found themselves thrown out of work overwhelmed the already slack labor market.[47]

The diversity as well as the skill and experience levels of many of the newly unemployed added to the problem. Choreographers and performers in the production show *Jubilee*, scheduled to open three weeks after the fire occurred, faced particular difficulties. Jobs in entertainment were especially scarce before the fire—and the sudden displacement of dozens of dancers, musicians, and showgirls, plus the production staff that made their performances possible—swamped the labor market. For cast members of *Jubilee*,

who had been in rehearsal for weeks, the fire was especially devastating. They had spent so much time at the MGM that they came to consider the resort home for the extended period the show was expected to run. "When I saw it on fire, I thought of it as my hotel, my home," one of them later explained.[48]

Janet Ford, one of *Jubilee*'s eleven principal dancers, began looking for any available job after the fire, even if it seemed monotonous and unchallenging. "You've got to make a living," Ford said. The displaced dancer also filed for unemployment benefits, and she received at least a few unemployment checks before finding a temporary job at the Tropicana Hotel, working as a "fill-in" showgirl. Most of Ford's fellow cast members had a more difficult time finding new employment. Some went north hoping to work in or around Reno. Others went out of state seeking jobs. "The family is being broken up," Ford lamented ruefully.[49]

The recollections of Jim Bonaventure, a showroom server at the MGM, offer insights into the plight of an altogether different category of displaced employees. With few prospects for a job in the aftermath of the fire, Bonaventure moved to Tonopah, a small town north of Las Vegas, where he found employment building homes. "There was no work right after the fire," he recalled, "so my wife called her aunt and uncle who were in Tonopah, who hired me to help frame houses." The move upended the lives of Bonaventure and his family. "Tonopah was a major change," he explained. "We lived in a 25-foot motor home, had no hook up to power for most of the first thirty days, and had to use propane to cook." His two sons remained in Las Vegas, staying with friends for nearly a year until conditions improved in Tonopah. "It was not a place for the boys to be at first," Bonaventure said. "After the first month things became more stable."[50]

As Bonaventure's experiences illustrate, many displaced workers assumed new personas, at least temporarily, as a result of sudden and drastic lifestyle changes after the fire. Those who entered new lines of employment had to adjust to new living conditions as well as new work routines. Even those who found jobs in or around Las Vegas that were similar to the ones they lost found themselves in new circumstances with new responsibilities, new coworkers, new supervisors, and new routines—even new working hours. Employees accustomed to sleeping at night sometimes now had to adjust to sleeping during the day, and vice versa. Accustomed to working at a large luxurious resort catering to upscale tourists, many MGM employees had to

take less exciting jobs in less glamorous places, where the clientele spent less money and left smaller tips.

The experiences of Kathy Griggs, a food server in the MGM coffee shop at the time of the fire, further illustrate this pattern. Griggs had three children and had recently separated from her husband. She found a new job shortly after the fire, but only because a coworker she ran into at the union hiring hall gave her own job referral slip to Griggs. "She knew I had kids and she said, 'you know what, they're going to have another job here, so you take this one,'" Griggs recalled. "I was surprised because we weren't like best friends or anything." The referral landed Griggs a job in the coffee shop at Sam's Town, a small resort on Boulder Highway. "At first I was just on-call," Griggs remembered. "I agreed to work whenever they needed me. But they kept scheduling me." Within weeks, Griggs was working full-time on the day shift. She appreciated the job, but her tips were smaller than at the MGM, and the work was "a lot harder." As she put it, "I had to do more manual cleaning at Sam's Town [and] more prep work." Griggs also spoke of a tense relationship with her new supervisor. "She was hard on me."[51]

Patty Jo Allsbrook, a switchboard operator at the MGM, offered another vivid example of the fire's impact on workers. It took Allsbrook several weeks to find a new job after the fire, and when she did find one, it was as a switchboard operator on the night shift at the Riverboat Casino, one of the Strip's smallest properties. "Someone recommended me to a supervisor there," she remembered, "and they called the Teamsters and requested me by name." Working nights changed the daily routines of Allsbrook and her family. "It's not for anybody normal," she said of her experience working the graveyard shift. "It's a whole new world." Her days off also changed from Saturday and Sunday to Tuesday and Wednesday, creating yet another disruption. "We couldn't do things as a family the way we used to," she remembered. "Tuesdays and Wednesdays became days for special celebrations." The changes forced Allbrook's husband to assume new responsibilities as a parent. He began taking their daughter to school, for example, and supervising her on weekends. "Everybody shifted modes," as Allsbrook put it.[52]

As these personal stories suggest, MGM employees were active agents in their own history and in this story of a disaster and its consequences. They found ways of helping themselves and each other, and had at least a few resources on which to fall back. Some had enough grit and initiative to leave

Las Vegas and find work in other industries. Many called on the resources of their families and friends to deal with the crisis and sought help from their union. The circumstances of displaced workers varied enormously. Some had spouses that were gainfully employed, and their own homes and savings accounts. Others were single parents who had to rely on workers' compensation and food stamps after the fire. The woman who gave Kathy Griggs her union work slip because Griggs was a single parent with dependent children is a marvelous story of worker solidarity and self-help. The story highlighted the complexities of the fire's aftermath and the fact that workers affected by it confronted problems in their own ways.

The MGM fire was a significant if temporary setback for Nevada's tourist economy. It raised the question of whether any of the state's numerous high-rise resorts were safe. Even a small grease fire, it seemed, could reduce the largest and most opulent resort to ruin and to the manifest danger of its guests. Public and industry officials acknowledged this problem somewhat grudgingly, while the state's leading newspapers suggested that would-be tourists were viewing its resorts as firetraps. "We do have an image problem," Democratic state senator William Hernstadt conceded, though the senator evidently regarded the problem as one of public relations rather than public safety. "The public perception is that some of our buildings are firetraps. It's wrong, and I tell members of the press that it's wrong, but still that is our national perception."[53]

But that perception was *not* wrong, as the MGM fire had grimly proved. In an effort to improve the image problem as much as to save lives, public leaders moved to address the issue directly by strengthening the state's fire codes for high-rise buildings. Governor List placed himself at the forefront of this movement. Shortly after the fire, after assuring the public that he supported fire safety reforms, List established the Commission on Fire Safety to inspect high-rise buildings throughout the state, assess their safety, and recommend necessary changes in the state's fire safety codes. Headed by the well-known banker (and future governor) Kenny Guinn, the commission immediately hired inspectors to conduct the necessary investigations, and it met eight times within the next three months to discuss the inspectors' findings and make recommendations. The panel concluded that Nevada already had some of the nation's most stringent fire codes, but then—to show how lax "stringent" meant in this context—recommended establishing a

state commission to evaluate procedures for preventing, containing, and extinguishing fires in high-rise buildings.[54]

As the panel completed its work, another high-rise fire underscored the urgency of the safety problem. On February 10, 1981, a troubled busboy with a history of committing petty crimes intentionally set fire to curtains on the eighth floor of the Las Vegas Hilton, a resort that rivaled the MGM in size and magnificence. What motivated the employee remains unclear, but he later said he was "messing" with drugs at the time, and had not meant to harm anyone. Regardless of his reason, the fire swept through the eighth floor of a hotel tower, killing eight people and injuring more than two hundred others. In a terrifying reenactment of the MGM disaster, tourists broke windows and leaned over balconies begging for help. Firefighters rushed into the resort and ascended stairwells, and helicopters hovered over the property to airlift people out of harm's way. The timing of this second disaster could not have been worse for the state and the industry. Reporters from across the country were in Las Vegas to cover Frank Sinatra's appearance before the state gaming board, which had recently expressed reservations about the popular entertainer's application for a gaming license. Newspapers everywhere published images of the Hilton fire and thus focused more attention on the problem of fire safety in Las Vegas.[55]

The Hilton fire marked a turning point in this story of a modern disaster and its consequences. This new tragedy offered clear, indisputable proof that the 1980 MGM fire was more than just an isolated incident or a freak accident. On the contrary, it showed that something terrifying—a monster of sorts—threatened a community and its tourist-based economy. This creature, or thing, left in its wake towering infernos and dead bodies. If there were any doubts about the danger that it posed, they now melted into air. From this point forward, there was no denying the fire safety problem. Now the general public saw a pattern, and the pattern had more emotional punch than a single event. The Hilton fire not only reinforced fears that Las Vegas resorts were unsafe but also reflected badly on public leaders and resort managers. It offered yet more evidence of policy failure, if not outright negligence. Disaster victims and their families certainly saw it that way. In the aftermath of these fires the victims filed an avalanche of lawsuits against the MGM and Hilton that ultimately cost the properties more than $250 million to settle.[56] As these risk bearers pursued the risk generators, they became active agents of change. Like the fire itself, the lawsuits drew considerable

media coverage and thus called more attention to the problem of fire safety. These lawsuits gave legislators and industry leaders more incentive to support fire safety reforms and all but obliterated opposition to it.[57]

As these scenes played out, Nevada finally gave the issue of fire safety the seriousness it deserved. On the heels of the Hilton fire, Democratic state senator Joe Neal of North Las Vegas introduced a sweeping fire safety bill that had been in the works for several weeks. The bill promised to make Nevada a leader in a national fire safety movement. It called for the installation of state-of-the-art sprinkler systems in all buildings over fifty-five feet in height and smoke detectors in all motel and hotel rooms. By the time legislators finished debating it, the bill required such buildings to have devices that shut down their heating and air conditioning systems during emergencies, plus exit corridors with emergency lights and doors that shut automatically. Elevators had to be fitted with devices that could automatically recall them to the first floor, and paging systems had to be installed in order to provide people with clear vocal instructions on how to safely evacuate a building. In addition, all areas in buildings used for public assembly had to have interior finishes made of fire-retardant materials.[58]

The legislation also called for the creation of a new regulatory agency—the Board of Fire Safety—to check architectural designs and inspect construction sites of proposed new buildings, and to ensure that they conformed to fire codes. It directed the board to evaluate new firefighting technologies and required that the most effective of those be incorporated into new buildings. Developers or building owners who wanted to use alternative technologies had to get the board's approval to do so. The board consisted of eleven members appointed by the governor, including a licensed architect, a general contractor, a professional engineer, three fire chiefs or fire marshals, and two representatives of the gaming and lodging industries.[59]

Such proposals drew strong support from trade unions, especially those whose members worked in the resort industry. Their concern was workplace safety. In urging passage of the legislation, Vern Balderston of the firefighter's union reminded legislators that lax building codes had caused working Americans untold harm. "When will there be a mandate to implement the installation of automatic fire protection sprinkler systems to stop these needless deaths?" Balderston asked. Sprinkler systems were proven to save lives. "There has never been a single fatality in a totally sprinklered high-rise structure." Areas of the MGM Grand that were fitted with sprinklers, Balder-

ston reminded lawmakers, suffered little damage during the 1980 fire. "I am convinced that if that building had been properly sprinklered we would not be discussing this now," he concluded.[60]

The resort fires in Las Vegas created a unique opportunity to strengthen Nevada's fire safety codes, but getting the powerful resort industry to support costly retrofitting programs required political maneuvering as well as a sense of duty and compassion. Senator Neal, the bill's chief sponsor, was up for the task. As chairman of the Human Resources Committee, which had jurisdiction over fire safety proposals, Neal was in a powerful position to push his agenda. To draft his legislation, he relied on the expert advice of State Fire Marshall Tom Huddleston, who understood how to prevent and fight resort fires as well as anyone. Only one other lawmaker agreed to co-sign Neal's bill, fellow Senator Hernstadt. Like Huddleston, however, Hernstadt proved to be a strong ally. He had previously owned and managed Las Vegas television station KVVU and knew how to promote policies and influence people. When Governor List presented a different version of the bill to the state assembly, Neal and Hernstadt incorporated parts of it into their own legislation and pushed forward with what came to be called the Retrofit Bill."[61]

Resort managers naturally worried about the costs of implementing this bill, but given the sensitivity of the situation and the obvious need to assure tourists of the safety of Nevada resorts, they never vehemently opposed it. They realized that large numbers of tourists might now cancel plans to visit Las Vegas, and that those who came to the city would not necessarily stay in a high-rise resort. Tourists could just as easily rent a nice motel room near the Strip, which suddenly seemed safer as well as easier on the pocket. Management wondered, too, about the impact on tourism in Reno and other parts of the state. Entrepreneurs like Kirk Kerkorian and Steve Wynn had big plans for developing Nevada, and so did national hotel chains like the Ramada and Hilton Corporations. Some of the older resorts along the Strip had recently launched new major construction projects, including the Dunes and the Sands. All of these plans now seemed problematic. Widespread television coverage of the two fires had threatened the state's entire resort industry.[62]

This safety movement took a major step forward in early March, when MGM executives announced plans to replace their old fire safety system with an advanced, computer-driven system capable of detecting and react-

ing to fires anywhere on their property instantly, on a twenty-four-hour basis. Designed by Johnson Controls of Milwaukee, Wisconsin, a pioneer in the field of air temperature control for nonresidential structures, the system would put sprinklers, smoke detectors, and public address speakers in every hotel room and suite as well as all other parts of the resort. It would include air-purging units that could quickly drive smoke out of any part of the property and a large computerized control room protected by firewalls. According to MGM President Alvin Benedict, the sophisticated safety system would make the MGM "one of the safest hotels in the world."[63]

The MGM's announcement put considerable pressure on other resorts in the state to follow suit, or at least to support the proposed retrofit ordinance. But reforms rarely materialized quickly or without extensive deliberation and debate. Acting largely through the Nevada Resort Association, other resorts agreed to support fire safety reforms but urged lawmakers to bear in mind the costs they were imposing on the business community. "One of the problems that any retrofit program has built into it is that there are costs imposed upon the owner without his consent," E. A. Schweitzer of the resort association told legislators. "This creates financial and functional problems that are generally unplanned and can have some very serious detrimental effects upon operating a business." "There is a business interruption function on any retrofit program which adds additional costs to the owner," Schweitzer continued. "There are times when your rooms cannot be occupied or there are areas of your food preparation that have to be separated, where you cannot function normally while you are in the process of construction."[64]

Robbins Cahill of the association also suggested that the legislation placed an unreasonable burden on property owners. A former chairman of the Nevada Gaming Control Board, Cahill reminded lawmakers that resorts had to obey not only the requirements of the new state safety code but also those of codes adopted by city and county governments. Those codes, the association feared, may be different, more demanding, and more expensive than the state code. How could resorts operate profitably, Cahill asked, if they had to keep redesigning buildings to satisfy both local officials and state regulatory agencies? Before requiring that existing resorts install new fire safety devices, Cahill urged state lawmakers to prohibit local governments from creating retrofitting specifications that differed in any way from those of the state. "There should be some sort of cap on the activities and the ordinances

of local governments," he suggested. "If we could just get some restriction of caps and some assurances as to what might happen to us in the future, I think this problem could be worked out."[65]

Such lobbying efforts resulted in a law that reflected the interests of both business and labor. Five months after the MGM fire, Governor List signed the Nevada Fire Safety Act. The act was a major step forward in fire safety. It both established the requirements and procedures already listed and insisted that Nevada's city and county fire codes be consistent with the state code. Any variation of that requirement, the act stipulated, had to be approved by the new Board of Fire Safety. From management's perspective, this provision gave a sense of stability to the business environment. It seemed practical and equitable, at least to large resorts that had the financial resources to invest in workplace safety. Once industry leaders threw their weight behind the legislation, it passed quickly.[66]

Because it was already under reconstruction, the MGM was the first major resort to fully meet the requirements of the law. As promised, when the MGM reopened on July 31, 1981, the resort had an impressive, state-of-the-art fire detection and control system. At the core of the system were two powerful computers, one of which functioned as a backup unit. Located in the security department behind fireproof walls, the computers helped security workers monitor scores of new life-saving devices, including more than thirty thousand heat-activated sprinkler heads. Two large pumps designed for transporting liquid supplied these sprinklers with enough water to prevent another catastrophic fire. They could push sixty to seventy gallons of water to individual guest rooms, each of which had at least four sprinklers.[67]

The system had many other components. A large electric generator protected the main computers against power outages, new exhaust fans could purge smoke from large areas within minutes, and the new air conditioning system prevented smoke from entering hotel rooms. A new public address system could provide guests with safety and evacuation instructions, and newly installed phone jacks near stairwells could help firefighters communicate with one another. Smoke detectors with audible alarms and manually operated fire alarms were positioned throughout the resort. When fire alarms sounded, stairwell fans increased air pressure to slow the spread of smoke, stairwell doors unlocked automatically, and elevators returned to the ground floor. The MGM even offered guests a short film about fire safety through its closed-circuit television system. Fire inspectors called the refurbished resort

the "safest hotel in the world." As Fire Chief Roy Parrish said of the resort's new safety features, "You name it, they've got it."[68]

The majority of the MGM's displaced employees, perhaps three-quarters or more, were back at work when the MGM reopened. "Everybody I knew came back," one of the returning employees later recalled.[69] Some had returned two or three months earlier, stocking kitchen pantries, installing slot machines, and otherwise preparing for the reopening. Choreographers, performers, and others involved in the production of *Jubilee* had also been back at work for some time, including musicians, wardrobe workers, and stagehands. Most employees who returned to the MGM did so at least in part to keep their seniority and the advantages that it offered, such as preferred work shifts, extended vacation time, and retirement benefits. Some of them had worked for years to earn positions that offered premium tips. Kathy Griggs had secured a counter position in the MGM coffee shop for that reason and returned to the position when the MGM reopened. She and other returning workers believed they had overcome a great challenge. "It's all behind us," one of them said of the dislocation. "This is the day we've been waiting for."[70]

The employees expressed confidence in the safety of their new work environment. Before reopening, the MGM gave many of its former employees tours of the new fire control system. Janet Ford, who returned to her job as a *Jubilee* dancer, learned detailed information about the new system. "We were given a briefing on the fire control computer," she told a reporter when the resort reopened. "It is an unreal computer system. We don't fear being in the hotel." Another employee, Kathy Jungels, had no hesitation about going back to work. "From what I've seen," she said of the fire control system, "it's absolutely fantastic."[71] Still another returning employee, Eileen Daleness, also saw this silver lining in the MGM tragedy. "We all felt very safe," she later recalled.[72] This feeling testified to the influence of tragic events on public policies as well as to the return of order and stability in the lives of many MGM employees. The drama had ended, its plot resolved.

Nevada's experience with resort fires motivated the state's lawmakers, employers, and labor leaders to join forces and address the issue of fire safety. The willingness of multiple parties to work together was understandable: media coverage of a towering inferno in a popular tourist destination had shocked the public. Television stations across the nation had broadcast ter-

rifying images of the MGM fire. The spectacle raised fundamental questions about the safety of all Nevada resorts and thus threatened the future of the state's single most important industry. Tax revenues and profit margins hinged on regaining the public's trust. The Hilton fire, which also played out on a nationally televised stage, exacerbated the problem. Like the MGM disaster, it meant expensive renovations, bad publicity, and costly lawsuits. In the wake of these tragedies, at a time when fire safety standards varied from state to state, and even city to city, Nevada crafted the nation's toughest fire safety laws.

What happened in Nevada had nationwide implications. By 1983, tourist-friendly Florida had followed Nevada's lead, mandating automatic sprinkler systems and smoke detectors in lodging establishments taller than three stories. Within the next few years, Connecticut, Massachusetts, and other states enacted their own retrofit ordinances, and several large cities adopted similar regulations, including Atlanta, Los Angeles, and San Francisco. Congress made its own contribution to this safety movement in 1990, when it passed the Hotel and Motel Fire Safety Act. Signed by President George H. W. Bush, the act prohibited federal employees on business trips within the United States from staying in multistoried lodging places that lacked sprinklers, and it charged the federal government with compiling and publishing lists of lodging properties that had installed these systems. The act thus gave property owners even more incentive to upgrade their fire safety systems, and it turned more public officials into fire safety advocates. By the end of the century, major hotel chains like Marriott, Ritz-Carlton, and Hilton boasted of their adherence to modern fire safety standards, and most American cities required sprinkler systems in new hotels and motels.[73]

This wave of new safety codes explains patterns that can only be described as progress. Between 1980 and 2015, the number of fires in American lodging establishments declined by approximately 70 percent. Even the worst of those fires claimed no more than fifteen lives, and the vast majority claimed none. Most did not spread beyond the rooms in which they started and were extinguished before fire trucks arrived on the scene.[74] This is not to say that the dangers of fires have been eliminated or that vigilance is no longer necessary. American firefighters still respond to some thirty-five hundred fires in hotels and motels annually, or nearly ten per day. Though the fires claim fewer than a dozen lives per year, they injure more than 120 people and cause roughly eighty-five million dollars in property damage. Eleva-

tor shafts and stairwells in many large hotels still allow clouds of smoke to travel upward through the properties, as do laundry chutes and other air ducts. Further, furniture and ornamental objects that feed fires are still scattered throughout many of these places. Even high-rise structures built according to strict building codes may be unsafe: sprinkler systems and other life-saving technologies have to be maintained, and evacuation plans must be effective. In short, fire safety remains an elusive goal.[75]

6

Federal Building Bombed in Oklahoma City!

> It could even be possible that the value of those good and honored things
> consists precisely in the fact that in an insidious way they are related to
> those bad seemingly opposite things, linked, knit together, even identical
> perhaps.
>
> Friedrich Nietzsche, *Beyond Good and Evil*

The moment is literally etched in stone, on twin bronze gates framing Oklahoma City's symbolic downtown memorial. At 9:02 a.m. on April 19, 1995, a large rental truck filled with fuel and ammonium nitrate exploded alongside the Alfred P. Murrah Federal Building. The blast threw an enormous cloud of dust over the entire downtown area, and by the time it settled, 168 people were dead and another 600 lay injured. Investigators soon learned that a former American soldier, Timothy McVeigh, had parked the vehicle next to the building before fleeing the area. An accomplice, Terry Nichols, had helped him assemble the "truck bomb" and also plant his getaway car. Both men saw themselves as modern-day crusaders fighting government tyrants and their minions, and they targeted the Murrah Building because it housed seventeen federal agencies, including those involved in recent raids on anti-government militia groups. Two years earlier, the men had vowed to avenge the deaths of seventy-five Branch Davidians who died outside Waco, Texas, after federal agents laid siege to the religious group's home and church.[1]

The Federal Bureau of Investigation (FBI) had McVeigh and Nichols in custody within two days of the bombing. The agency's first clue appeared on day one, when it discovered the rear axle of a truck near the bombsite. A

partial serial number on the battered object led federal agents to the Ryder Rental Corporation, which matched the numbers to a truck rented at an agency in Junction City, Kansas. There, the agents pieced together a composite drawing of McVeigh and then learned that he had stayed at a nearby motel. Though McVeigh used an alias in renting the truck, he provided the motel with his real name as well as the former Michigan address of his accomplice. Once the agents had McVeigh's name, they discovered he had just been arrested in Perry, Oklahoma, for driving a car without license plates and carrying a concealed weapon. By April 22, when agents found bomb-making materials in Nichols's new residence in central Kansas, they knew they had their men. A grand jury indicted the men for conspiring to blow up a federal building and kill its occupants, and they were eventually convicted of the crime. McVeigh received the death penalty for his role in the crime and died by lethal injection in 2001. Nichols received life imprisonment for his actions, without the possibility of parole.[2]

The Oklahoma City bombing has attracted the attention of a wide variety of authors. Journalists and popular writers have uncovered the bombing's most intriguing details. Some have written about McVeigh's fascination with firearms and foreign policy, for example, and his disdain for the federal government. Others have focused on the somber, more tragic aspects of the bombing, like the fact that fifteen children perished in the Murrah Building's second-floor day-care center, and that rescue workers had to amputate a young woman's leg to free her from the rubble.[3] Social scientists have shown how churches and charitable organizations responded to the bombing, and analyzed the ways in which the media interpreted the event. They have also placed the bombing within the context of rightwing militia movements and domestic terrorism, and explained how the tragedy affected national politics. Legal scholars have debated the merits of the antiterrorism legislation that the tragedy inspired. That legislation, they pointed out, insulated state death penalty sentences from oversight by federal courts.[4]

Yet not much is known about the Murrah Building itself or the men and women who worked within it. Before the bomb destroyed it, residents of Oklahoma City generally viewed the Murrah Building with a great sense of pride. When it opened in 1977, the building was a sign of Oklahoma's emergence as a modern Sunbelt state. With its long, glass-paneled northern facade, the building looked like an ultramodern structure. But investigators concluded the building was an easy target for terrorists. The blast from the

Rescue workers attend a memorial service outside the bombed-out Alfred P. Murrah Building, May 5, 1995. The Oklahoma City bombing killed 168 people, more than two-thirds of whom worked in the Murrah Building. Associated Press

bomb shattered the structure's immense glass face, sending thousands of razor-sharp glass fragments flying in every direction. It also lifted the building's upper floors off their supporting columns, triggering a progressive collapse of the northern walls.

At that moment, 361 people were inside, almost 300 of whom were federal employees. Nearly 30 other people in the building worked for a privately owned credit union, and 3 were employed by the day-care center. Sadly, 117 of these working people died that morning, accounting for more than two-thirds of all fatalities resulting from the explosion.[5] Workers who survived the bombing may have considered themselves fortunate, but their future was anything but bright. Many not only suffered debilitating physical injuries

but also showed signs of post-traumatic stress disorder. Moreover, the bombing had completely destroyed their work environment and thus obliterated their normal routines and relationships. There were no plans in place for relocating these survivors or for helping them cope with the loss of friends and coworkers. The 1995 tragedy, then, was both America's worst act of domestic terrorism and one of the nation's worst workplace disasters. As such, it prompted public officials to find new ways of protecting federal buildings.

The origins of the Murrah Building lay largely in the early 1940s, when wartime spending finally lifted Oklahomans out of the Great Depression. At the beginning of World War II, the US Department of Defense viewed oil-rich central Oklahoma as the perfect place to build large military installations and wartime production facilities. The area's normally mild climate, clear skies, and flat terrain made it particularly suitable for testing large aircraft and training pilots, and its remoteness from the nation's coastlines also made it less vulnerable to potential air attacks. With strong support from Oklahoma's congressional delegation, the federal government financed the construction of Tinker Air Force Base, about ten miles east of downtown Oklahoma City, as well as a nearby aircraft manufacturing plant. Those two facilities were among several federally funded structures that sprang up on the Oklahoma Plains during the war and early postwar years.[6]

The rise of these facilities set in motion a chain of events with far-reaching consequences. To begin with, the facilities' operations created tens of thousands of jobs. The new airbase near Oklahoma City alone generated nearly fifteen thousand jobs, and the aircraft plant another twenty-three thousand. The rise of the new facilities expanded opportunities in all sectors of the local economy. The air base increasingly resembled a small city whose inhabitants required a steady supply of food, water, and electricity. The aircraft factory, like all defense plants, required the constant delivery of oil, processed minerals, and machine parts. The demand for goods and services had enormous implications for real estate, banking, and transportation. It spawned new warehouses, grocery stores, schools, and much more. Indeed, wartime spending in Oklahoma created the conditions for a quarter century of growth and prosperity.[7]

Postwar developments in the state's capital city underlined the dynamic at work. When the war began, about 220,000 people lived in greater Oklahoma City. Though the city's boundaries covered only fifty square miles,

more than 90 percent of local residents lived within those boundaries. In 1970, however, some 640,000 people called the larger metropolitan area their home. By then, the city had annexed many of its surrounding communities and covered an area of six hundred square miles; nonetheless, only 60 percent of the local population lived within the city's limits. The growth of neighboring towns had spiked sharply, making the greater metropolis one of the busiest places in the Sunbelt West. The increasing number of commercial airliners at the city's main airport mirrored the new reality. Located only six miles southwest of the downtown area, the Will Rogers Airport, a former military airfield, serviced all of the nation's major airlines. Those airlines offered daily nonstop flights to cities across the country.[8]

Road improvements in the postwar years proved especially significant to Oklahoma City. In 1947, Oklahoma authorized the construction of a major turnpike connecting its capital city to Tulsa, the state's second-largest metropolitan area. A decade later, it approved plans for an additional turnpike to connect the capital to the growing town of Lawson, Oklahoma. Meanwhile, the federal government's new interstate highway system put Oklahoma City at the crosshairs of two larger thoroughfares: Interstates 35 and 40. On road maps, the city began to look like the hub of a great wagon wheel, with spokes spanning outward in every direction. Big freight trucks and small cars alike traveled the roads in greater and greater numbers, bringing more and more goods and inhabitants to central Oklahoma.[9]

These developments easily justified the expansion of Oklahoma City's federal facilities. In the early 1970s, to the delight of civic leaders, the federal government contracted a local architectural firm to build a new federal complex in the downtown area, where the government already maintained a large courthouse and post office. The firm, founded by its chief architect Wendell Locke, proposed a rectangular-shaped nine-story structure made of reinforced concrete, with enough interior space to house regional offices for more than a dozen federal agencies. The proposal included a four-story subterranean parking structure whose top floor featured a parklike plaza with fountains and plants that could serve as a playground for children as well as a relaxing place for employees to take their breaks. Federal officials, acting largely through the Government Services Administration (GSA), agreed to pay the firm fourteen million dollars to build the complex.[10]

On October 21, 1974, a small but enthusiastic crowd of some forty people attended a groundbreaking ceremony for the new complex. The event was

well planned and full of pageantry. Uniformed high school band members performed music as the people arrived, ushers and usherettes seated the crowd and passed out programs, and honor guards from the armed forces marched through the area carrying colorful flags. As the ceremonies began, prominent public leaders ascended a small wooden stage framed by red, white, and blue bunting. One by one, they addressed the crowd from a podium draped with government seals. After the city's mayor made a few opening remarks, a local Baptist minister delivered an inspiring invocation, and three US congressmen praised the planned complex. As the congressmen proudly noted, the soon-to-be-built structure would bear the name of Alfred P. Murrah, a distinguished federal judge whose career in Oklahoma City had spanned three decades.[11]

Before thrusting their shovels into the ground, the congressmen stressed that the people of Oklahoma needed easier access to federal services. Senator Henry Bellmon, a Republican serving his first term in Congress, complained that citizens had been "wandering all over town" in search of government agencies. Jokingly, the conservative congressman said the Murrah Building would be a better place to solve problems than Washington, DC. "I'm not suggesting this building be the beginning of the relocation of the nation's capital to Oklahoma City," he kidded. Representative John Jarman, a Democrat serving in Congress since 1951, linked the planned structure to democracy and good governance. "This building represents a significant step forward in the federal government's efforts to be more responsive to the needs of the people," he concluded. Tom Sneed, a House Democrat since 1948, shared that view. According to Sneed, the Murrah Building marked "one big step forward towards underwriting a great future for Oklahoma City."[12]

An architectural sketch of the Murrah Building rested on an easel next to the makeshift stage, highlighting the structure's modern, energy-efficient design. Large expanses of glass stretched across the building's long northern walls to maximize natural light. Smaller shaded windows poked out of its south side, which was more exposed to the afternoon sun. The main entrance faced south and opened onto the landscaped plaza atop the easily accessible parking structure. The first two floors along the north side would be slightly recessed from the rest of the building, providing a bit of shade in addition to extra space between street traffic and the rear entrance. Four

concrete columns stood proudly within that space, adding support for the upper floors.[13]

When it opened three years later, the Murrah Building met all existing government building codes. The structure had an ordinary concrete frame with large, circular tube columns at each of its four corners. Buried within the third floor, a wide concrete beam supported additional columns that helped carry the weight of the higher floors. Thin panels of stone and gypsum plaster separated offices within individual floors, and gridworks of lightweight materials hung from each floor's ceiling to conceal unsightly pipes, wiring, and ductwork. The building's flat roof included synthetic rubber and other insulated materials that helped shield the building's occupants against the extreme weather conditions of the Great Plains. Unlike most structures in the area, the building featured a state-of-the-art fire protection system with numerous indoor sprinkler heads and voice-activated alarms. These and other safety features gave occupants an extra sense of security.[14]

The GSA had the responsibility of operating and maintaining federal structures, and it had a noticeable presence inside the Murrah Building. Like most large property management companies, GSA employed budget analysts and contract specialists along with maintenance mechanics and custodians. At the time, the agency already managed more than a dozen federal structures in Oklahoma as well as a large amount of leased space within privately owned buildings. It had a regional office in the capital city's federal courthouse, where its employees monitored the life cycle of government structures by testing and checking inventory items and supervising the work of contractors who renovated or cleaned the buildings. GSA employees also collected rent, provided office supplies, and made sure that property taxes, insurance premiums, and other bills were paid on time.[15]

The GSA made all the space assignments within the Murrah Building and thus determined exactly where workers would perform their duties. Those assignments generally reflected the specific needs and concerns of federal agencies and their employees. The GSA placed the US Postal Service on the ground floor, for example, where postal workers set up a small mail-distribution center near a loading zone. It put the Social Security Administration (SSA) at ground level, too, making it easier for elderly and disabled citizens to visit the agency. The top floor went to the US Secret Service, the Drug Enforcement Agency (DEA), and the Bureau of Alcohol, Tobacco, and Fire-

arms (ATF), whose employees often worked on sensitive investigations and needed extra privacy and protection. With the GSA's help, the three law enforcement agencies reinforced the top floor's hallways with special, heavy metal dividers and set up a security checkpoint to keep unauthorized individuals from wandering through the offices.[16]

In its effort to maintain the building, the GSA gave its tenants a long list of "good house-keeping rules" that were prefaced with a personal plea to its occupants: "GSA is proud of this new Federal Building and its facilities. It will require our joint efforts to maintain it in good condition and pleasing appearance." The first of these rules concerned office walls. "Locate furniture and equipment in such a manner as to minimize the marring of walls," it read. "Avoid the use of tacks or tape on walls," another said. If tenants wanted to hang plaques or pictures, they were instructed to contact a GSA representative: "He will be pleased to help you." Most of the rules encouraged cleanliness: "Both for sanitary and safety reasons, immediately clean up any spilled food or beverage." "Put trash in containers" and "keep restrooms clean," ordered other rules. The final rules encouraged tenants to conserve energy. "Exercise economy in the use of electric current, heat and other services," they instructed.[17]

The responsibility for protecting the Murrah Building and its occupants lay with the Federal Protective Service (FPS), which maintained a small office in the nearby Federal Courthouse. Formed in 1971, the FPS provided law enforcement and security services to all federally owned and leased buildings. In its early years, FPS officers patrolled the Murrah Building and monitored surveillance cameras placed in and around the building from a control center in the courthouse. Budget cuts, however, gradually undermined the agency's ability to provide those services. By 1995 the Murrah Building's security system had been reduced to one roving guard who focused on keeping vagrants out of the building's lobby area and thieves out of its parking garage. The camera surveillance system no longer functioned well. People could enter and leave the building as they pleased, without passing any security officers or working surveillance cameras.[18]

To put it plainly, the Murrah Building was low-hanging fruit to terrorists like McVeigh and Nichols. The building's architectural design, though impressive, was simply out of touch with the times. By 1995, terrorist acts at home and abroad had grown more deadly, and some federal buildings were far more vulnerable to attack than others. Prescient public officials might

have envisioned such a tragedy. Only two years earlier, terrorists detonated a truck bomb in New York City, beneath the north tower of the World Trade Center. Though the bombing killed only six people, it might well have led to the tower's complete structural failure, as the perpetrators had intended. But the real harbinger of doom occurred in 1983, when terrorists in war-torn Beirut, Lebanon, used truck bombs to attack two large buildings that housed an international peacekeeping force. The attack killed some 300 people, including 220 US marines. When Congress investigated the bombings, the public learned that the marines had died in a four-story structure built on flat ground along a busy street, which made the place harder to defend.[19] In the wake of the tragedy the State Department established a blue-ribbon advisory panel that recommended stronger security standards for all overseas embassies. Among other things, the panel concluded that American embassies should be set back from public streets. If embassies could not meet new safety standards, the panel added, they should be replaced. In hindsight, those standards should have applied to federal buildings on American soil.[20]

The Murrah Building's workforce was relatively well educated, diverse, and stable. About half of its federal employees had college or professional degrees, and almost all had at least a high school education. Most had several years of work experience in their chosen field, either in public service or the private sector. There were as many women as men in this workforce, and minority groups were well represented in most job categories. The salaries of these employees varied according to their skills and work experience. Though their pay generally lagged behind that of their counterparts in the private sector, employees at the Murrah Building enjoyed relatively good job benefits, including paid vacations and family-friendly leave policies together with health insurance and pension plans. Supervisors could and sometimes did discharge federal workers for poor performance, but jobs were relatively secure. Many of the employees belonged to the American Federation of Government Employees (AFGE), a national labor organization whose contracts with the government protected the rights and interests of its members.[21]

Daily activities within the Murrah Building revealed the importance of federal jobs. Patterns and routines in the large Department of Housing and Urban Development (HUD) were one example. Established in 1965 as part of President Lyndon B. Johnson's Great Society programs, HUD was a multifaceted organization that employed 125 people on the building's seventh

and eighth floors. Some of these employees collected data on Oklahoma's housing market and promoted local projects that made housing more affordable. Others worked to enforce federal laws that prohibited discrimination against disabled citizens and minority groups, or managed grants that assisted elderly and disabled Americans in living independently rather than having to move into nursing homes or hospitals. Still others promoted the construction of homes on the land inhabited by Oklahoma's many Native American groups.[22]

Federal employees in other agencies made equally important contributions to the local community. At the Federal Highway Administration (FHA), located on the fourth floor, thirty employees helped to maintain the vital interstate highway system. This meant investigating local highway accidents and planning how to mitigate traffic congestion. It also involved evaluating new road construction projects, determining whether contractors adhered to federal construction standards, and reviewing the costs of those projects. At the Department of Agriculture, located on the fifth floor directly above the FHA, fifteen employees helped local farmers and ranchers solve common problems. They collected and distributed information about new methods for producing crops and provided training on how to inspect local meat and poultry products. These employees also combated problems such as invasive pests and animal diseases, protected local water supplies, and helped farmers and ranchers acquire bank loans.[23] In the Veterans Administration (VA), also located on the fifth floor, about a dozen employees helped veterans make the transition from military to civilian life. The VA staff included a counseling psychologist, two case managers, two home loan appraisers, and two administrative assistants.[24]

In addition to its federal tenants, the Murrah Building housed two private organizations, the Federal Employees Credit Union (FECU) and America's Kids Child Development Center. The FECU employed thirty-three people on the building's third floor to provide financial services to federal employees and their relatives. Its workforce included financial advisors and loan officers as well as accountants, clerks, and tellers.[25] The childcare center was just below the credit union, in an area that stretched along the north side of the building near the street. On the morning of the bombing, the center's three employees were caring for twenty-two children under the age of six, most of whom belonged to two-career families or single working mothers employed in the downtown area. Some of the children were apparently near

the second-floor windows, only a few feet from where the truck bomb exploded. The owner of the small business, a local Sunday school teacher who was out of town at the time, later told the press that the children often stood by the windows looking out at the world around them. "[They] loved to watch the clouds, to watch the cars, to watch the people pass by," she said. People outside, the owner added, enjoyed seeing the children, too. "People always liked to walk by, on the sidewalk, and look at the babies there."[26]

When the bomb exploded, the federal employees in the building had just begun their morning work routines. Many were at their desks reading memos, analyzing data, or talking on the telephone. Some were standing at counters assisting their morning clients or walking through hallways on errands. A few were in the fourth-floor snack bar on the south side of the building or in the large storage room directly above it. Whatever their circumstances, the explosion killed or injured almost all of them. According to the Oklahoma State Department of Health, the blast killed 45 percent of the building's occupants and injured another 47 percent. In other words, it killed or injured nine out of ten people in the building.[27]

All of the building's federal agencies and its two private companies suffered terrific losses that fateful morning. The HUD, the largest of the federal tenants, lost 35 of its 125 employees, and the SSA lost 16 of its 61 workers. A few smaller agencies lost more than a third of their respective workforces, including the Department of Agriculture, the Department of Defense, and the FHA. The credit union suffered an even worse blow, losing 18 of its 33 employees. The toll the blast took crossed race, gender, and ethnic lines. Old timers and newcomers died. Top administrators and highly trained specialists fell alongside receptionists and clerks. The only people who survived the disaster without suffering serious injuries had been along the building's southern and western walls, which remained largely intact.[28]

Survivors vividly remembered the moment of impact. Lorri McNiven of the SSA, for example, was training a new employee. "We had discussed her first appointment for the day," McNiven later recalled. "I remember her collecting her papers from my desk, and then she took them to her desk." That was the last McNiven ever saw of the new employee. When the bomb went off, McNiven initially went into a trancelike state. "I remember a feeling like I was falling, then I remember thinking I was dying, that I had suffered an aneurysm, stroke, or heart attack, or maybe I had electrocuted myself from having my hands on my computer." She then lost consciousness for nearly

an hour. When she regained consciousness, she started "feeling around." "All I could feel were metal edges, cement, glass and grit." The blast had left McNiven barefoot atop a row of file cabinets and covered with rubble. One of her coworkers eventually found her there and helped her out of the building. As she soon learned, the explosion killed more than a dozen of her co-workers along with two dozen visitors who had been in the agency's large reception area.[29]

Captain Henderson Baker, an operations officer in the army's fourth-floor recruiting office, had a similar experience. Baker was standing at a co-worker's desk discussing work-related matters when everything suddenly changed. "We talked for about five minutes and then I started to fall," the officer remembered. "At first I thought this was an earthquake, then I thought I'm dreaming, and then I thought I'm dying." Baker remained conscious as he fell to the ground floor. "It felt like thirty minutes because of all the things that went through my mind," he said. "When I hit the floor I was dazed, but I got up." Through the dust and darkness, Baker saw a "ray of sun" that he followed to the street. Looking back at the Murrah Building, the officer realized he had survived something horrific. Despite his own injuries, Baker spent the next few hours assisting with the search and rescue efforts. "I thought of the other soldiers and civilians that were in the office. I immediately went back into the building to see if they were OK." He soon learned that two of his coworkers as well as five civilians in the recruiting office had died.[30]

HUD employee Donald Bewley had an equally traumatic experience. Bewley had just shown a coworker how to retrieve financial statements using her computer when the bomb exploded. "I don't remember hearing the blast, only feeling a large gust of wind," he later recalled. "I do remember the ceiling tile and everything above the tile coming down on me—the sprinkler system, power line, phone lines and concrete." The explosion pushed Bewley into what he called the "airplane crash position." "The blast did not remove me from the chair I was sitting in. It only secured me there." After the initial shock, the HUD employee struggled to make sense of what had happened. "I remember standing up only to see nothing but concrete dust, followed by heavy black smoke from the cars on fire across the street." Bewley suddenly realized that most of the seventh and eighth floors—where HUD was housed—were gone, and that large file cabinets blocked his only escape route. Bleeding from deep head wounds, Bewley managed to crawl to a nearby stairwell. "From there

it was a matter of walking down the stairs." When he reached the street, a stranger drove Bewley to the South Community Hospital, where doctors stitched up his wounds.[31]

The story of Priscilla Salyers, an investigative agent for the US Customs Service, is equally poignant. Salyers was sorting through her mail on the fifth floor when the bomb exploded. A senior special agent had just walked over to her desk, located about twenty feet from the windows along the northern side of the building. "I glanced up and he said something to me, but I couldn't hear him because there was this noise, like a sonic boom," she recalled. "Our eyes locked, and then everything went black." The blast completely disoriented Salyers. "I felt as though I were being thrown around while sitting in my chair, and felt that I was being forced forward and going to hit my head on my desk." The customs agent then heard a "wind-like noise" and saw several flashes of light. "I remember thinking I might be having a heart attack or some seizure." Salyers found herself lying face down and unable to move. "All I could feel was rubble, metal bars, and big cement boulders on my head." Thinking she was still in her office, Salyers was expecting a coworker to help her up when she heard someone say, "we have a lot of children in here!" The sad truth was that the agent had fallen three floors into the childcare center on the second floor and was buried so deep in rubble that rescuers could not extract her for nearly four hours. When they finally did, Salyers learned she had three broken ribs, a punctured lung, and multiple abrasions. She spent the next six days in treatment at St. Anthony's Hospital.[32]

Florence Rogers, the chief executive officer of the privately owned credit union in the Murrah Building, had a compelling story of her own. Rogers had scheduled an eight thirty meeting in her office with seven staff members, and they were midway through it when the CEO's world changed forever. "We covered a few of the items and I had just turned around from looking at the next agenda item on my computer screen at 9:02 when suddenly the building literally blew up before my eyes, taking all of the staff members meeting with me down, with six floors from above falling on top of them," she later explained.[33] Miraculously, Rogers had been seated atop a concrete slab that stayed intact during the explosion. The concussive force of the blast knocked the CEO to the floor, but she remained in a small portion of the building that never collapsed. Bleeding and disoriented, Rogers initially wondered where her coworkers had gone. "Where did you all go?" she called out.

Rogers then came to her senses and screamed for help. "I gathered myself up and started yelling out the window that was no longer there," she recalled. Within minutes, two surviving employees came to her rescue. When Rogers finally reached the street level, she "began to hyperventilate," and then "the shock set in." Rogers learned that more than half of her staff was missing and presumed dead. Five other credit union workers had suffered debilitating injuries, including one with a doorstop lodged in her head.[34]

As these stories demonstrate, injuries among survivors varied widely. Hospital records suggest that more than half of all survivors suffered wounds to their head, face, or neck. Half also had eye or auditory injuries, and roughly a third had lacerated tendons and nerves. Yet another third suffered injuries to their arms and legs, which typically meant broken bones or bone dislocations. A large but undetermined percentage also inhaled so much smoke and dust that they damaged their respiratory systems. Recovering from these injuries often required several months—or even years—of physical therapy and counseling.[35]

The case of HUD employee Clifford Cagle shows how extensive some injuries could be. Cagle was at his desk when the bomb exploded. "I was knocked off my chair by the force of the blast," he later recalled. Chunks of concrete and glass hit Cagle's face and neck, crushing the left side of his skull and shredding his left eye. By the time rescue workers got him to Presbyterian Hospital, Cagle had lost nearly five pints of blood, and his heart had stopped twice. When doctors found that glass had penetrated the membrane between Cagle's skull and brain, they initially feared he had suffered serious brain damage. Over the next year, surgeons gradually reconstructed Cagle's skull, rebuilt his eye socket, and fitted him with an artificial eye. The process required eight separate surgeries and involved breaking Cagle's jawbone, pulling eight teeth, and reconnecting many nerves. Between surgeries, he had to wear a mask that doctors modified every other day and braces to straighten his teeth.[36]

GSA employee Tom Hall also suffered debilitating injuries. Hall was standing in a coworker's office on the building's ground floor when the bomb went off. "All of a sudden I felt like I was being electrocuted," Hall recalled. "I felt a hot vibration going through me." The blast threw Hall to the ground and left him covered with concrete and fiberglass. When Hall tried to get up, he realized that his left leg had been crushed. "The pain was so great that I

would pass out and then awaken seconds later," he remembered. Hall's memories of being rushed to the University of Oklahoma Hospital put a finer point on the seriousness of his injuries. "Just like in the movies they raced me into the emergency room with all the bright lights and put the mask over my mouth and nose and I was out." It took surgeons several hours to assess all of Hall's injuries, which included a severed jugular vein as well as a cracked left femur and serious nerve damage. He spent ten days in the hospital before returning home. An orthopedic surgeon and a physical therapist gradually helped him regain most of his strength and mobility.[37]

Recovering from injuries required help from family and friends, even for survivors with relatively minor wounds. Looking back on her own experience, Dianne Dooley, a vocational specialist for the Department of Veteran Affairs, highlighted the point. The bombing shattered Dooley's right wrist, index finger, and thumb and thus kept her from carrying out many basic daily routines, including dressing, cooking, and bathing. "I had no use of my hand," she explained. For weeks after the bombing, Dooley depended on her husband and in-laws for support. Her husband regularly changed Dooley's bandages to clean her wounds, gave her pain pills every few hours, and paid all the monthly bills. "My husband had to be a nurse," as she put it. Dooley's neighbors also played a role in the recovery process. "The house became a revolving door," she recalled. "Neighbors were bringing food, cooking, and even cleaning."[38]

Healing emotionally from the bombing was as challenging as recovering from physical injuries. Even survivors with minor injuries had seen gut-wrenching, unforgettable images that tragic day, such as the dead bodies of the children in the day-care center. Many had lost close friends in the bombing and witnessed the tragedy's devastating impact on their friends' families. "We mourned not only for the loss of great friends, but also for their families," one of them explained.[39] Their own close brush with death left emotional scars and even a sense of guilt. Many wondered why had their own lives been spared. Several people survived only because they did something out of the ordinary that day—like taking an early break. Tom Hall, for example, had stopped to chat with a coworker after leaving a business meeting. "I would have been at my desk next to the glass and only twenty feet from the Ryder truck and would surely have been killed," he noted.[40] A handful of employees survived the bombing because walls, floors, and office furniture collapsed around them in an odd, particular manner. GSA employee Kathy

Brady was one of the fortunate few. "I truly believe that what saved me from much more serious injury was that my overhead systems furniture storage cabinet fell over on me, acting like a mini-bomb shelter," she concluded.[41]

About 60 percent of the surviving employees sought counseling services after the bombing. Most suffered from some form of post-traumatic stress disorder (PTSD). Symptoms of PTSD include severe headaches, sleeplessness, and depression as well as loss of appetite, memory lapses, or lack of concentration. Some survivors had all of these symptoms. "For me, depression set in and life went out the door. I could no longer sleep, eat, or function as a mother and wife, daughter, sister or friend," one of them explained.[42] "I couldn't even load the dishwasher," another said. "Things just didn't fit into place."[43] As one professional counselor who treated survivors later explained, symptoms of psychological disorders often appeared long after the bombing, but patients did not always connect them to the tragedy. "Feelings that we deny or bury sometimes resurface years later," the counselor said. His own patients included spouses and children of the deceased and also several emergency responders at the Murrah Building that day.[44]

Quite a few survivors found comfort and hope in their religious beliefs. Evangelical Protestantism has long been a central feature of Oklahoma's culture, and many of the men and women who worked in the Murrah Building were part of that community. A few of these individuals believed that the Bible was inerrant in its claims about the greatness of God and his Son, Jesus, but most embraced a more modernist, or "mainline," theology that saw the Bible as a general guide for dealing with the challenges of life. They tended to view their personal problems as part of larger spiritual challenges, and thus the bombing presented itself as a solemn test of their faith. Some no doubt drew strength from the story of Job, the biblical figure who lost his children, health, and property—the devil's way of testing his faith—without blaming God for the tragedies. Like Job, these men and women maintained their faith in the worst of times. They believed that God had protected them when the Murrah Building collapsed and that he would continue to protect them in the future.

Personal stories again illustrate the point. Consider the words of survivor Ruth Schwab, who worked for HUD. Like so many employees in the building, Schwab was at her desk when the bomb exploded. "Suddenly, there was a tremendous boom. Then it seemed like I was being hurled down a dark, black hole," Schwab recalled. "I couldn't see anything." Rescue workers

later pulled Schwab from the rubble and rushed her to the hospital, where she underwent eight hours of surgery to remove shards of glass from her face and neck. Though she lost her right eye that day, Schwab maintained her religious conviction. "God pulled me through," she later declared. "God saved me physically, emotionally, and spiritually. That's the only way I can explain it."[45]

The words of HUD employees Rhonda Griffin and Sharilee Lyons put a fine point on this perspective. Griffin, who suffered her own serious injuries during the bombing, confessed to having an epiphany while recovering from her wounds. "The Lord spoke to my heart about three weeks after the bombing, when I thought I couldn't make it any longer," she believed. "If it weren't for God, I wouldn't be here today to tell my story."[46] Lyons came to a similar conclusion. At the time of the bombing, Lyons happened to be in Washington, DC, with three of her coworkers attending a computer course. She interpreted that fact as some form of divine intervention. "God put his protective shield around me and kept me safe," she surmised.[47]

Yet the bombing left other employees wondering if God really existed, or whether he was as powerful or benevolent as they had previously believed. Dot Hill, a GSA purchasing agent, was among those employees. Hill, who nearly suffocated to death before getting out of the building, had real emotional wounds after the disaster. "My world fell apart," as she put it. "Death became my strongest desire, my only true companion." Hill wondered if the Devil himself had outsmarted God that day, and whether he had saved her only to make her suffer. "My rational thoughts were gone and so suffer I did, with every breath I took." But Hill ultimately regained her faith and relied on it to recover. "Without God I was not going to make it," she realized. Recovering from the experience nonetheless proved difficult. "It took 18 months for me to begin getting a part of myself back, to care about my family, friends, myself."[48]

As always, great tragedy presented conundrums for religious leaders. People asked yet again, How could a strong God allow so many innocent people to die? Had God's purpose really prevailed? Evangelist Billy Graham, regarded as the most influential American preacher of modern times, addressed such questions a week after the bombing in a memorial prayer service held in a large multipurpose arena at Oklahoma City's State Fairgrounds. Speaking to more than ten thousand people in the arena and to many more watching on television, Graham said he had often been asked why a God of

love and mercy permitted such terrible misfortunes. "I don't know," Graham admitted. "I have to confess that I never fully understand—even for my own satisfaction." Graham recalled walking through the ruins of earthquakes, hurricanes, and typhoons, and standing at the bedsides of dying children, wondering aloud, "Lord, Why?" "There is a mystery to it," Graham counseled. "There's something about evil we will never fully understand this side of eternity." Like so many church leaders in Oklahoma had already done, Graham assured his audience that God loved them and even shared in their suffering. "Some of you today are going through heartache and grief so intense that you wonder if it will ever go away," he recognized. "I want to tell you that our God cares for you and for your family and for your city."[49] But even beloved preachers armed with the most inspirational biblical passages failed to alleviate the widespread pain and suffering. Indeed, as a previously popular song had asked, "How can you mend a broken heart?"[50]

Regardless of their faith, survivors attended scores of funerals and made numerous hospital visits after the bombing, which sometimes heightened their own sense of loss and loneliness. Jennifer Takagi, a funding specialist at HUD, went to funerals for three weeks after the event. "I went to 15 to 18 funerals," she recalled. "Sometimes I went to two a day." The funerals occasionally left Takagi in a state of despair. "The level of depression was so great," she explained. "Everyone was so emotionally damaged."[51] But the funerals also helped survivors cope with the tragedy. As Beverly Rankin of the Social Security Administration explained, the funerals offered a strange sense of comfort. "We needed to be together," Rankin said. "There was a lot of crying, a lot of hugging."[52]

As those words reveal, the bombing united everyone who survived it. Within the Murrah Building's workforce, survivors like Rankin forgot about previous complaints or feelings of resentment and began to see their displaced coworkers in an entirely new light. While struggling to overcome their own grief and injuries, they also tried to help one another. Sociologist Charles Fritz, a pioneer in the field of disaster studies, called attention to this phenomenon years earlier, when he described disasters as "unifying forces." In making his point, Fritz referenced human behavior after earthquakes, maritime accidents, and aerial bombing campaigns. Time after time, he explained, disasters strengthened bonds between groups and produced a sense of "mutual aid and solidarity."[53]

↓

Resuming work seemed impossible at the time. The Murrah Building was beyond repair, and no other structure in the area could house all of the displaced agencies. Most of the agencies had lost their top managers and supervisors along with critical documents and files, including records that helped them keep track of their clients and financial transactions. Countless decisions had to be made: Where could the agencies relocate? Who would lead the effort to reopen them? How would deceased employees be replaced? What guidelines governed the treatment and compensation of the injured survivors?

The GSA took the lead in answering these questions. GSA administrators from across the country rushed to Oklahoma City after the bombing and set up a command center a few blocks south of the bombsite, at the four-hundred-room Medallion Hotel.[54] The administrators divided themselves into several working groups to assess losses and plan for the future. The latter task involved helping the people in charge of displaced agencies locate new office space in the city and rebuild their workforces. GSA employees who survived the bombing provided essential information at the command center. These employees explained the layout of utility lines in the downtown area, for example, and identified the administrative needs of particular agencies.

A former GSA maintenance mechanic, Richard Williams, was particularly helpful. Williams had worked in the Murrah Building since it had opened in 1977 and knew the operational aspects of the structure as well as anyone. "I knew that building like the back of my hand," he later noted. Though seriously injured by the building's collapse, Williams provided vital details about the building and its federal tenants. His role in the rebuilding process grew larger during the next few weeks, when he was finally healthy enough to go to the command center. Williams's initial impression of the center revealed the complexity of the process. "When I made my first visit to the hotel after about two weeks, I was amazed at the talent and level of expertise being used to solve all the problems related to the bombing," he recalled. "There were various group meetings in progress. All had flow charts and information all over their meeting rooms."[55]

It took a herculean effort to reestablish some federal agencies, especially HUD. Immediately after the bombing, HUD's top administrators set up their own communications center at Oklahoma City's downtown Public Housing Office. Three weeks later, after sifting through tattered records and adding

up their losses, the administrators established a small, temporary office in an old Montgomery Ward store a few blocks from the bombsite. There, a small staff of survivors and new "detail employees" who arrived from distant HUD offices handled a variety of pressing matters. The staff processed life insurance and worker compensation claims, for example, and began hiring new workers to replace those it had lost. The agency eventually leased two floors of the old store and completely renovated the space. By the time it opened its doors to the public in July, most of its surviving employees had returned to work.[56]

The memories of Germaine Johnston, one of HUD's managers in the Murrah Building, shed more light on the challenges facing displaced agencies and their surviving employees. Johnston was fortunate on the day of the bombing in that "nothing heavy fell on me." Despite a few minor injuries and the shock of losing so many friends and coworkers, Johnston reported to HUD's downtown communication center almost every day after the bombing, where she helped evaluate the agency's losses and reestablish services. "I could not stay home. I needed to be with the people who shared the bombing experience." Johnston recalled the local telephone company setting up a dozen phones in the center that she and the new "detailees" used to communicate with surviving employees. "We had charts on the walls on flip-chart pads with lists of employees," Johnston said. Meanwhile, real estate contracts had to be signed, construction sites inspected, and other issues resolved. "During all of this, our customers and clients were clamoring to have their needs met."[57]

As HUD's survivors trickled back to work, they adjusted to new circumstances. They were required to learn how to operate a variety of new office machines, including new computers, printers, and fax machines. "Everything had to be relearned," Johnston recalled. "We had very little assistance on most of the machines, just booklet after booklet to read." But HUD did offer its employees help with using new computers and software. It contracted with a private company that offered seminars designed to help workers accomplish specific tasks on computers. "We were sent to their facilities to take day-long, sometimes even week-long, hands-on classes," Johnston said. HUD also contracted with a psychiatrist from the Veterans Administration to provide counseling sessions for survivors. "Each employee had to meet with him for at least fifteen minutes, at least every two weeks."[58]

Other agencies also provided expert counseling for their surviving em-

ployees. The SSA, for example, which reopened in a shopping mall north of the downtown area, brought in therapists from the nearby Indian Health Service to meet with survivors on an individual and group basis. The counselors not only helped survivors cope with their emotions but also briefed new employees on how to work with trauma victims. It was all part of what the agency called the "long-term healing process."[59] The effectiveness of those counseling sessions was not immediately apparent, however. Several months after the bombing, many survivors still had short attention spans and long memory lapses. They remained easily frightened and were often frustrated and fatigued.[60]

The recollections of former social security workers who survived the blast underscore the difficulties of reestablishing normal routines. Janet Beck said she had "memory problems" or concentration lapses "for months after returning to work." As a claims representative, Beck had interviewed several people a day before the bombing, in face-to-face meetings or over the telephone. But interviewing people after the bombing proved unexpectedly difficult, even by telephone. "I was scared to death to take the calls," Beck remembered. "I was worried I couldn't handle it." The fact that social security claimants wanted to talk about the disaster made the job all the more difficult. "People wanted to hear our story," Beck explained.[61] Beverly Rankin, who also fielded calls from social security claimants, said she had trouble accomplishing even the most basic tasks long after the bombing. "I wasn't worth anything for four to six months," as she worded it. When clients inquired about the disaster, Beck said she had trouble holding back her emotions. "People would come in and ask about the bombing. It triggered tears."[62]

Reestablishing the privately owned credit union was especially challenging. Unlike the Murrah Building's federal tenants, the FECU had been housed entirely in the now blown-out structure. Fortunately, two of CEO Rogers's chief assistants were out of town at the time of the bombing and were thus able to help save the devastated firm. More than a dozen local credit unions helped, too. The companies sent several of their own employees to work with Rogers and also provided much needed office supplies and equipment. One of them even provided the use of its recently built corporate headquarters at Tinker Air Force Base, where Rogers and her staff spent nearly two months before relocating to an office building in the suburban community of Bethany, located about ten miles northwest of the downtown area. Though the FECU lost almost all of its assets, including tens of thou-

sands of dollars in checks and cash in the bombing, the company stored many of its financial records at an offsite location, and data processing specialists and telecommunications experts helped retrieve that information.[63]

As Rogers later recalled, rebuilding the FECU required both sacrifice and staying power. "The days, weeks and months following the bombing were long, chaotic and busy ones." Despite suffering cracked vertebrae and many cuts and bruises, Rogers began working at the company's new airbase headquarters only a few days after the disaster. She remembered the new workplace vividly. Phones rang constantly, news reporters showed up unannounced, and there were customers everywhere. The place itself looked more like a funeral home than a credit union. Flowers covered tables and desks, and strangers were often crying and hugging one another. At a time when there was so much to do, Rogers found herself shorthanded. "There were only 9 of my initial 32 staff able to report back to work to begin the rebuilding process." Even those employees had difficulty carrying out normal routines. Rogers described them as "crippled both physically and emotionally." In the initial weeks after the bombing, Rogers spent twelve to thirteen hours a day at work, dealing with insurance agents, workers' compensation issues, grieving employees, and much more. But she spent most of her time interviewing replacement workers, a process that lasted for several weeks. "Hiring replacements for those lost was a tremendous challenge that we thought would never end," the CEO explained.[64]

In short, getting back to work was a slow, complicated, and trying process. It meant making scores of seemingly inconsequential yet crucial decisions, like finding new routes to the workplace, new places to eat lunch, and new office machines. It also meant working with new people, including supervisors who directed daily activities and appraised work performances. As Jennifer Takagi of HUD later noted, the bombing created "new lines" at work. Survivors like Takagi saw themselves as members of a distinct group, yet there were divisions even among the survivors. Some employees recovered from the disaster much more quickly than others, both emotionally and physically. A few survivors, Takagi said, "just couldn't get past that day."[65]

These words play up the problems and challenges that employees faced after such traumatic experiences and in doing so tell us more about the true costs of disasters in work environments. These tragedies amounted to more than just scenes of bloodied bodies and collapsed buildings. The number of people they killed and injured has never been the only yardstick available for

measuring their impact, nor have the costs associated with shutting down operations, replacing lost buildings, or settling lawsuits. As the personal recollections from Oklahoma City show, these were life-changing events that demanded displays of courage and resilience. They challenged people's sense of identity as well as their faith in the future and in some cases even kept them from functioning as normal human beings. By upending the presumed stable order of things, they also made the world look all the more confusing and threatening.

Public officials promised to replace the Murrah Building quickly, but doing so took much longer than anticipated. The process began immediately after the bombing, when President Bill Clinton directed the Justice Department to evaluate the safety of the eight thousand facilities owned or leased by the federal government and to recommend new minimum-security standards to reduce their vulnerability. To carry out the president's directive, the department formed two groups of security specialists—one to inspect national facilities and the other to explore new ways of protecting them—and the groups made their recommendations in the summer of 1995. In effect, their proposals called for new security standards at all federal structures as well as significant changes in the design of future federal facilities. To implement those recommendations, the president soon established a new agency within the executive branch—the Interagency Security Committee (ISC). The ISC included representatives from all executive branch agencies along with several security experts appointed by the president.[66]

Congress generally applauded the president's action, even though conservatives consistently railed against the power of "big government." For years, champions of small government had described public workers as overpaid "paper pushers" if not "slackers and moochers." They also suggested that the productivity of federal workers paled in comparison to their counterparts in the private sector. When the Murrah Building fell, however, the halls of Congress echoed with tributes to federal employees. Spokespeople from both sides of the political divide described federal workers as heroes and patriots, and even suggested they outperformed government workers around the world. That was certainly the perspective of Democratic Congressman Tom Foglietta of Pennsylvania. "The American civil servant is perhaps the best civil servant in the world," Foglietta told the House shortly after the bombing. "All one has to do is travel abroad to see that American

federal employees are second to none in terms of their devotion to the job," he added.[67] For weeks after the bombing, Congress discussed proposals to protect federal workers, as if their lives were in grave danger. "The safety of our Government workers and their children hang in the balance," Republican Congressman Jack Quinn of New York said during one of these discussions. Never before had federal workers been viewed as so important or appreciated.[68]

As the situation unfolded, the GSA looked for a place to build a new facility in Oklahoma City. The process took far longer than anticipated. Public officials in Oklahoma campaigned for a new national memorial on the land where the Murrah Building had stood, and the GSA could not easily purchase alternative sites in the downtown area. The agency initially hoped to establish three separate structures in the area but could not acquire a key parcel of land needed for the project. It ultimately settled for a single structure adjacent to the proposed memorial on two city blocks between the business district and an old downtown neighborhood. To design the structure, the GSA commissioned a Chicago firm whose principal architect, Carol Ross Barney, had studied the reconstruction of European cities after World War II. Barney proposed a long, horseshoe-shaped structure that construction crews began building in 2001. The complex opened to the public a little more than two years later.[69]

Though not an impenetrable fortress, the new Oklahoma City Federal Building was no doubt the nation's safest federal facility. The structure stood on a gently sloping site in a relatively quiet part of the downtown area. Its west wing was only two stories high, and the east wing was just one story taller. Its outer walls were unusually thick and made of dense, "poured-in-place" concrete that was exceptionally durable. The concrete used to construct floor slabs and ceilings fastened directly onto large support columns that extended into the structure's foundation, and steel bars spiraled through the columns in unique ways to keep the building from collapsing under extreme pressure. The building was set back fifty feet from the closest street and surrounded by an assortment of solid objects, including a stone wall and granite boulders as well as several concrete bollards and metal posts. Parking was restricted to a lot northwest of the structure that held seventy-six cars, and the closest drop-off lane was more than twenty feet from the main entrance.

Additional safety features also loomed large. Big steel plates reinforced

The Oklahoma City Federal Building, 2005. The building's exterior walls, columns, bollards, posts, and suspended roof are among its many security features. Photograph by Steve Hall, courtesy of Ross Barney Architects

the building's main entrance, which opened into a spacious lobby with soaring ceilings and thick concrete walls. Before anyone could enter the lobby, armed guards checked their identification and personal items and made them walk past sensitive x-ray machines and magnetometers. Many of the building's doors were bulletproof, and all of its windows had steel frames and two panes of tempered glass designed to stay within the frames in the event of an explosion. Surveillance cameras were everywhere, from the outer edge of the parking lot to interior hallways and offices. No one could approach, enter, or walk through the building undetected.[70]

Public officials held an official dedication ceremony in the building's U-shaped courtyard on May 3, 2004. About six hundred people attended the event, including many of the Murrah Building's surviving employees. Attorney General John Ashcroft, who provided the ceremony's keynote address, recognized the gravity of the moment. "It is an honor for me to be here to

Backside of the new Oklahoma City Federal Building, 2005. The rear walls, boulders, columns, and rock garden offer additional protection against terrorist attacks. Photograph by Steve Hall, courtesy of Ross Barney Architects

share this day of hope and this day of healing," Ashcroft told the crowd, adding, "You've lived in the aftermath of man-made destruction." Ashcroft delivered a personal message from President George W. Bush to the survivors of the tragedy. "By building in the face of loss," the president's message read, "we honor both those who have survived and those who died that day." From the president's perspective, the new federal facility was a "proud symbol of perseverance."[71]

By the time of the ceremony, nearly a dozen federal agencies had already moved into the new structure, and a few more were in the process of doing so. Not all of those agencies had previously been housed in the Murrah Building, however. As it turned out, many of the federal workers who survived the 1995 bombing opposed moving to the new complex. Those who had grown comfortable in their current locations generally resisted moving back to the downtown area, saying they could best serve their constituents by remaining where they were. At the Social Security Administration, em-

Inside the Oklahoma City Federal Building, 2005. Surveillance cameras, blast-resistant walls, bulletproof doors, tempered glass, and security guards protect the structure's main entrance. Overhead skywalks connect its two separate wings. Photograph by Steve Hall, courtesy of Ross Barney Architects

ployees argued convincingly that the agency's elderly constituents would have trouble getting downtown and walking the distance from the building's parking lot to its front entrance. Beverly Rankin was among these employees. "Most of us did not want to go back downtown," she recalled.[72] The agency's administrators ultimately decided to remain at the shopping center north of downtown.

More than a few survivors flat-out refused to make the move, including

Dot Hill of the GSA. "I said from the get-go I wouldn't go," Hill explained.[73] Hill justified her opposition by saying she could not look at the new federal complex without thinking about the bombing, but she also feared that terrorists might one day attack the new structure. The building's hardened structure and safety features, she believed, "would only force terrorists to be more creative."[74] Other surviving employees objected to working so close to a memorial that reminded them of the friends and coworkers they lost at the site. "They didn't want to look at that every day," Richard Williams of the GSA said.[75] As such words suggest, moving into the new federal building represented yet another challenge for survivors, especially those who had the most trouble recovering from the bombing.

Federal agencies that did move into the new complex made special accommodations for their surviving employees. They allowed the employees to work at alternative locations, for example, or helped them find jobs in a different federal agency. The largest agency that moved into the new complex, HUD, gave its survivors the option of remaining in the old Montgomery Ward store. "They told us we would not have to go to the new building," Jennifer Takagi explained. HUD's regional administrators also gave the survivors a tour of the new federal complex before it opened. The tour convinced several of the survivors that the place was not only safe but also more comfortable than where they presently were. The administrators even allowed employees to select their workstations within HUD's assigned space. "People got to choose where to sit," Takagi explained. Unlike many of her coworkers who moved into the complex, Takagi chose to sit near a window.[76]

Survivors who made the move were generally glad they did so; Dianne Dooley of the VA was one of them. After recovering from her injuries, Dooley returned to work at the old downtown post office, where her displaced agency had opened its local offices. Dooley grew frustrated in that work environment because the VA had to scatter its employees throughout the building. "I had to run down the hall to coordinate daily activities," she complained. The new federal complex brought VA workers into a common area again and also put them in larger offices with new furnishings. Dooley described the new federal complex as "modern," "sophisticated," and "convenient." She liked her new office and appreciated the large ground-floor break room where she could have lunch. She was also fond of the open courtyard patterned after Native American ceremonial grounds. Filled with rocks, bollards, and

heavily reinforced columns, the courtyard not only looked artistic but also offered protection against terrorist attacks.[77]

As surviving employees like Dooley walked through the new federal complex, many came to realize that safety had not been a high priority in their old work environment. "In the Murrah Building people were free to come and go," Dooley explained. She had often seen suspicious characters in the downtown area, some of whom were experiencing homelessness or mental illness. The fact that Dooley worked with a number of troubled war veterans added to her concerns. "I was always a little bit leery," she admitted. The presence of armed guards and surveillance technology at the main entrance was reassuring. As Dooley said, "I felt safer because I knew they were going to have guards in the new federal building, and they were going to have a screening process for people to have to go through."[78]

Other employees felt safer, too. Jennifer Takagi said security in the new federal building had helped her put memories of the bombing behind her. It comforted Takagi to see security officers patrolling the parking lot and inspecting bags at the main entrance. Even longtime employees had to show special identification badges when they entered the building. "You hold your badge up or you do not get in," Takagi explained. "I forgot my badge one day and I got patted down like everyone else," she joked. "They recognized me, I had my driver's license, but I left my I.D. at home." On some days guards patted down all employees as they entered the building, regardless of whether they had identification badges. "It keeps us on our toes," Takagi confessed. The security measures were so thorough that Takagi believed she would never witness another major disaster, at least at work. "I believe that with all my heart," she said.[79]

As time moved on, survivors recognized that the bombing had altered their lives in the most fundamental ways. Cynthia Gonyea of the US Customs Office said it best: "Everything has changed now." "We all have to begin anew."[80] The disaster affected workers' basic attitudes about their friends, family, and everyday life, and it even shook their confidence in their faith. Nonetheless, most bomb survivors eventually saw a bright side to the tragedy. John Gabriella of HUD, for example, said the bombing made him a better person. "I'm a whole lot more patient than I used to be," he explained.[81] GSA accountant Steve Pruitt, who suffered significant hearing and vision loss as a result of the bombing, said the tragedy taught him to appreciate

everyday life. "I've gotten a taste of how rough life can really get," Pruitt said. "I appreciate being alive, having my wife, two healthy sons and living in a nice house."[82] Social Security worker Liz Thomas developed a similar perspective. "You never know when your last day on Earth is going to be," Thomas warned. "With family, friends, co-workers—mend your fences before it's too late."[83]

The Oklahoma City bombing jolted the public consciousness. More than any previous disaster in recent memory, it reminded Americans that they lived and worked in a risk-filled society, where large and seemingly safe structures can suddenly come crashing down. The event differed from other disasters that destroyed large places of employment in two obvious ways. First, its perpetrators were not employers bent on maximizing profits or machines gone wild. This was no accident but rather a carefully planned crime, carried out by two homegrown terrorists. Second, most of the men and women affected by this tragedy were federal employees, an occupational group that had rarely been associated with disasters of any kind.[84]

The destruction of the Murrah Building needs to be seen as more than a terrorist attack, however, and as more than a crisis for any single occupational group. Its place in history is much larger than that. The bombing was also part of a panorama of workplace disasters that had shattered the lives and dreams of Americans for more than two centuries. Many of these disasters drew widespread attention to matters of public concern and thus hastened the rise of new public safety policies. The U-shaped structure that replaced the Murrah Building mirrored these patterns and trends. When it finally opened in the new millennium, public officials described the structure as America's safest federal building. Its castle-like defenses spoke poignantly to two interrelated themes of historical concern: how disasters destroyed modern work sites and how nations have confronted the anxieties and insecurities that these tragedies induced.

Conclusion

There is much to be grateful for . . . There are also many things to deplore, and to correct.

Kofi Annan, *We the Peoples*

Disasters have wrecked work sites throughout American history, in all parts of the nation and all sectors of the economy. The most notorious of these events have given rise to a powerful narrative that still shapes our collective memory. Like gruesome horror films, the narrative has its own sinister symbols, ranging from smoke-filled mineshafts and burning factories to horse-drawn ambulances and small wooden coffins. Its story lines have played out in older industrial centers of the late nineteenth and early twentieth centuries, usually in the urban Northeast or in mining communities of rural Appalachia. Paradoxically, the tragedies that fit this mold have often had cathartic endings. On many occasions, large-scale catastrophes have generated tidal waves of reforms, and the responses to disasters have come to be seen as potent forces of progress.

This book encourages readers to think about such events in new ways. Though each case study tells a separate story, the stories when assembled form a narrative of their own. These studies explore five catastrophes that shocked the nation in the half century after World War II, a time when service-oriented industries became the nation's leading engines of job growth. The five disasters unfurled far from the cradle of the Industrial Revolution, in a section of the country that was still relatively undeveloped before the war began. They destroyed places of employment that seemed safe compared to the nation's old factories and mines, and they affected a relatively wide

range of working people, including highly trained, salaried professionals as well as blue- and white-collar groups. But these tragedies also spelled disaster for scores of customers and other "nonworkers" who had the misfortune of being in or near the ill-fated environments. The toll they took on the general public suggested that almost *everyone* was in danger of being killed or injured and thus put added pressure on public officials to prevent similar tragedies in the future. These closing pages offer an opportunity to flesh out the nuances of this modern narrative and to reflect more carefully on its meaning and historical relevance.

To start this journey, we need to remember that the disasters discussed in this book were not minor mishaps or ordinary accidents, but rather incredibly destructive events that caused widespread disruption and physical harm. Their cruel plots began when some life-threatening danger emerged in the realm of work. They then lurched forward, at different rates of speed, toward the moment of impact. At that point, alarm and fear swept through the environment. People showed concern for others around them, but escape was the only chance for survival. The plots ground to a halt in their final, *post-impact* stage—a time of evaluation, recuperation, and reform. By then, the tragic events had become energizing, galvanizing forces, strong enough to change the way the nation organized and managed business and work environments.[1]

Each of these events raised basic questions about the effectiveness of public safety policies. The first study, which investigated the destruction of the Monsanto Chemical Plant on the Texas coast in 1947, showed that neither private businesses nor the federal government had adequately safeguarded the public against hazardous materials like fertilizer-grade ammonium nitrate (FGAN). The second study, of the tragic air accident over the Grand Canyon in 1956, suggested that business and government had failed to manage the dangers of air travel. Similarly, the powerful earthquake that rocked the San Fernando Valley in 1971 revealed the weaknesses of seismic safety regulations, and the MGM Grand Hotel fire of 1980 exposed the shortcomings of fire safety standards. The Oklahoma City bombing laid bare the public's vulnerability to terrorist attacks.

Arranged chronologically, our case studies trace rapid urban and industrial developments in the American West during and after World War II. To reiterate a central point, America's entrance into the war had profound con-

sequences for the nation's southwestern cities and states. During the war's early stages, the federal government financed an array of defense-related projects in areas of the country where the climate and terrain favored steady, year-round production. That ushered in an age of spectacular growth from Texas and Oklahoma to Southern California. Western historians have generally associated this growth with regional progress and national triumph, and understandably so. By turbocharging the economy, mobilization made the greater Southwest a more modern, prosperous place while all but assuring military success. By the late twentieth century, the region had become the Sunbelt West, an affluent, multicentered section of the country that influenced all aspects of American life, from politics and economics to literature and music.[2] As a few historians have emphasized, however, this "transformation" laid the groundwork for many new problems and challenges, from water shortages to environmental degradation.[3]

The studies shed light on the transformation's unintended consequences. The tragedy in Texas City offered the clearest example of how wartime spending set the stage for postwar calamities. The federal government had financed the construction of Monsanto's Texas City plant during the war to boost production of much-needed synthetic rubber. In the early postwar years, when the city's piers and warehouses overflowed with explosive cargo, the plant's volatile raw materials increased the potential for a major catastrophe. The Grand Canyon tragedy also needs to be read against the war's backdrop. Wartime spending had lured millions of Americans to Southern California, thus leading to the area's surging population in the early postwar years. These developments easily justified the expansion of the Los Angeles airport in the late 1940s. By the mid-1950s, large fleets of commercial planes lined the airport's runways and terminal facilities, which connected Southern California to the wider world. It was more than a coincidence, then, that in 1956 two commercial airliners took off from the airport only minutes apart, or that the planes had similar flight paths.[4]

The other disasters should also be viewed in the context of dramatic urban and regional growth. The San Fernando Valley earthquake would obviously have been less costly had it occurred before the war began. By 1970, the valley's rich farmland had given way to scores of new roads, homes, and businesses. The population of Sylmar, the community hit hardest by the quake, jumped from thirty-five hundred to forty thousand, and the number of structures in the community—including medical facilities—rose accord-

ingly.[5] Even the MGM fire and Oklahoma City bombing need to be understood in the context of wartime developments. Both Nevada and Oklahoma benefited handsomely from wartime spending. Military bases, magnesium factories, and aircraft manufacturing were among the many signs of federal investments in the states. These investments set in motion patterns that warranted the construction of more and more private businesses and public buildings. When the MGM Grand and the Murrah Federal Building opened in the mid-1970s, their shiny exteriors mirrored the emergence of two prosperous and expanding Sunbelt communities that had been "liberated" from their prewar past.[6]

Yet these disasters also need to be viewed outside of their regional context and within a broader, universal story about the side effects of modern industrial and technological advances. Each of the events draws attention to the close relationship between forces of creation and destruction, and to hazards of the modern world. Social theorists such as Ulrich Beck and Anthony Giddens have taken a keen interest in modern catastrophes, suggesting that they reflect the increasingly dangerous nature of contemporary life. Echoing the words of Henry George, the nineteenth-century American who famously said that "the 'tramp' comes with the locomotive," these theorists have reminded us that advanced societies produce their own endangerment.[7] "In advanced modernity," to quote Beck, "the social production of wealth is systematically accompanied by the social production of risks."[8] These risks may be clearer in some geographical areas than others, but they can be found throughout the modern world. As the risks grow more conspicuous, they not only cause grief and suffering but also expose the self-endangering nature of modern life, or what Giddens called "the double-edged character of modernity."[9]

The technical advances that help to explain our disasters are many-sided and somewhat obscure. The development of FGAN lay behind two of the tragedies. This chemically juiced, granular material proved to be a low-cost, all-purpose fertilizer, but it was also inherently dangerous. The ruins in Texas City made clear that the material had to be stored and handled carefully, and those in Oklahoma City showed that it might explode by way of a detonator and fuel oil. Technical innovations in the field of aviation lay beneath the Grand Canyon tragedy. In 1956 the commercial airline industry was still a new area of enterprise. High-flying planes capable of carrying dozens of passengers across the country without refueling had only recently

been developed. Equipped with air-cooled engines, air pressurization systems, and other new devices, aircraft like Lockheed's "Super Connie" could take travelers to heights where humans struggle to breathe, at speeds that posed their own risks.[10]

The march of invention also helps to account for the tragedy on the Las Vegas Strip. In the early postwar years, technical advances in structural engineering encouraged the rise of multipurpose, multitowered resorts, which literally elevated the dangers of hotel stays. The introduction of synthetic materials in hotel furnishings created additional hazards. When the MGM's ground-floor furniture and ornaments caught fire, they emitted deadly toxic smoke that climbed through the resort's elevators, laundry chutes, and other air ducts to its higher floors. New technologies cannot explain the California earthquake, but it bears repeating that nature alone did not account for the quake's devastating toll. Most of the structures the quake destroyed had not been built to withstand the energy released by such powerful temblors. While the number of homes and businesses in the Los Angeles area soared, concerns of seismic safety advocates had largely been ignored. In short, all of these disasters were bound up in a combination of regional, industrial, and technological developments. This "progress" gave rise to what Beck called our "risk society."[11]

Matters of victimhood and villainy in these studies have proven to be revealing on multiple levels. The working people affected by these disasters were not miners, garment workers, or other familiar groups of the distant past, but rather people who held commonly recognized jobs of our own era. Their personal backgrounds and skill levels varied widely. Some had little formal education and relatively few skills, such as day laborers, kitchen workers, and housekeepers. Others had advanced degrees and impressive skills, among them chemists, engineers, and airline pilots. Still others were well-trained public servants, including nurses, clerks, and office administrators. The incomes of these workers reflected this diversity. Some of the affected workers earned relatively high annual salaries, yet many others worked for minimal hourly wages. Women and minorities were well represented in these cases. In the ruins of the Murrah Building, women made up more than half of the deceased employees.

The survivors of these tragedies had important stories to tell. Their recollections have helped to show workplace disasters as more than brief

moments in time when structures suddenly burned or collapsed. They reminded us that the significance of these events cannot be measured by body counts and cleanup costs alone. For large numbers of working people, these events marked the beginning of long periods of recovery, reconstruction, and even relocation. They often forced independent, hardworking citizens to rely on the charity and goodwill of others, including their families and friends. We have always defined workers' struggles as part of an age-old, unavoidable conflict between capital and labor. But some of the greatest struggles in labor history had nothing to do with job-oriented issues like wages and work routines, or the right to form trade unions. They centered instead on how workers dealt with the unexpected destruction of their work environments and, again by implication, their physical injuries, lost income, mortgaged properties, and retirement programs. The meaningfulness of work has always been hard to measure. To many disaster survivors, however, work stood at the center of their personal universe. It held things together, including their sense of identity and self-worth. With the exception of the 1956 midair collision, which no one survived, these disasters forced working people to adjust to all sorts of new circumstances, some of which proved overwhelming.[12]

This is not to imply that the survivors of workplace disasters are disengaged victims, incapable of meeting the dilemmas they faced. On the contrary, many were active and resilient. Those who escaped serious injuries often played vital roles in relief and recovery efforts. When buildings collapsed at Sylmar's Veterans Hospital, shaken doctors and nurses set up triage units and treated wounds. Meanwhile, security guards ran through the surrounding community looking for emergency assistance, and other employees dug through rubble searching for survivors. When the MGM Grand caught fire, a housekeeper raced through the resort warning guests to evacuate, a lifeguard assisted paramedics, and pit bosses secured the casino's cash. Surviving employees contributed to the recovery process long after first responders had left the scene. In Texas City, these employees assisted in clearing ground for reconstruction projects and in evaluating the condition of damaged rooms, tools, and equipment. In Oklahoma City, they provided investigators with vital information about the layout of the fallen federal building and identified pressing problems facing their agencies as they rebuilt. If disasters pushed communities through periods of ruin and

resurrection, these instances of workers helping others were vital to the process.

The toll these tragedies took on groups of nonworkers—such as airline passengers, hospital patients, and tourists—proved informative, too. Perhaps most importantly, it underscored new realities in American labor markets. By the mid-twentieth century, the nation's service sector had already begun generating more jobs and wealth than did mining and manufacturing industries. But this trend gathered speed throughout the postwar period. In small towns and big cities alike, job seekers found growing opportunities in the areas of health care and tourism. Retail stores, real estate companies, and other service-oriented businesses were also hiring. Job opportunities were also expanding in the public sector, at all levels of government. The rise of this postindustrial age grew more apparent as nations of Europe and Asia recovered from the war. In the 1960s, American manufacturers began to lose their advantage over foreign competitors, forcing many companies to lay off large numbers of workers. By the 1970s, an age characterized by baby boomers and the Beatles, the service sector reigned supreme. Consumers appeared to be everywhere—in shopping malls, amusement parks, downtown offices, and countless other places. Long lines at checkout counters spoke volumes of the forces reshaping the economy. When disasters struck these places, they understandably affected more consumers than people on the job. Yet that should not obscure the impact of these disasters on workers and their places of employment.

The villains in these stories were not simply greedy employers more concerned about matters of cost efficiency and profit than the well-being of their employees. Unlike mining accidents and factory fires of earlier times, these disasters cannot be attributed to the exploitation and oppression of workers, or to the degradation of work routines and rising levels of inequality. Aside from the case of Oklahoma City, the issue of villainy in these tragedies was relatively complex and puzzling. Investigators never figured out exactly why the FGAN-laden *Grandcamp* exploded in Texas City, although they agreed that outdated methods of handling the ship's cargo were partly to blame. Pilot error and cloud coverage may well have contributed to the 1956 air tragedy, but airlines had stubbornly adhered to old principles of flying, and the nation's air management system was woefully outmoded at the time. MGM executives obviously should have installed a better fire safety

system when the resort was under construction, but there was nothing especially unusual about their decision not to do so. Investigators eventually attributed the fire to faulty electrical circuitry that overheated near the casino, though the design of the giant resort clearly fed the fire. Even the disaster in Oklahoma City reflected more than pure evil. The Alfred P. Murrah Building's glass-covered walls and proximity to the street made it an easy, convenient target for terrorists. Though the building's occupants felt safe, they were all too vulnerable to the dangers of an increasingly polarized society.

Each of these disasters had enough *focusing power* to create opportunities for legislative and organizational reforms. Nonetheless, it took a host of individuals and groups to force policy change.[13] The most important agents of change were elected officials who held powerful positions in the policy-making process. These figures had enough clout, credibility, and political savvy to broker deals in policy communities, where advocacy groups worked diligently to protect their separate interests. The nation's presidents have often played crucial roles in setting post-disaster agendas, and that was true in these cases. Presidents from Harry Truman to Bill Clinton literally left their signatures on paths to recovery. So did elected officials in the US Congress, directors of federal agencies, and state governors. Though none of these actors controlled the policy-making process, they pledged to make work sites safer and through their privileged positions made policy reform a matter of top priority. They had access to institutional resources that proved critical to public safety, such as disaster relief funds, emergency equipment and supplies, and classified information about national security.[14]

Actors at lower levels of government often wielded outsized power in the agenda-setting process. Following the 1971 earthquake, for example, county supervisor Kenneth Hahn and state senator Alfred Alquist proved to be *policy entrepreneurs*. When the quake hit, both officials had a reputation for promoting public safety issues, including seismic safety. With a wealth of connections and skills, they developed policy proposals and convinced businesses and public agencies to adopt and implement them. State senator Joe Neal proved equally influential in Nevada's fire safety movement. He too had good negotiating skills and was well situated in the decision-making process. Looking back on his role in the movement, Neal acknowledged the importance of his position: "I pushed it through the Legislature mainly be-

cause I was chairman of the Human Resources Committee, which took care of all fire legislation."[15]

To accomplish their goals, however, these various policy brokers relied on the advice of credentialed experts. Many of these experts were themselves leaders in their communities and had grown accustomed to working with public officials in committees and subcommittees. They also had connections with larger networks of professional organizations and private businesses, and were willing to sacrifice their time and energy for the common good. Their contributions to the process of standard setting were not completely selfless, of course. Cooperating with policymakers boosted the experts' visibility within their professions and no doubt helped them hone their technical skills. Some of the experts were themselves public servants, such as Nevada's fire marshals and specialists within the US Army Corps of Engineers. Others were scientists and academics, like Charles Richter and Clarence Allen, who advised policymakers on establishing new seismic safety standards. Still others worked for private businesses, such as Carol Ross Barney, the architect who designed the Oklahoma City Federal Building.[16]

The recurring appearance of trained professionals in these stories highlighted the relevance of technical advances to safety reforms. Indeed, reforms generally hinged on the availability of new technologies. Making the skies safer after the Grand Canyon disaster required the introduction of new radar equipment and better communication systems. Reinforcing hospitals in California depended on the use of seismographs, soil sampling devices, and other new tools. Workplace safety in Nevada resorts required the installation of new smoke detectors, automated sprinkler systems, and better public address systems. In Oklahoma City's new federal building, safety meant deploying new x-ray machines, surveillance cameras, and tempered glass as well as new bulletproof doors and super-hardened walls. Yet reforms also involved using old tools and technologies in new ways. Defending Oklahoma's new federal building, for instance, involved the deployment of stone walls, steel poles, and a rock garden. Old or new, these devices and defenses were icons of the modern historical narrative. They showed how we "moderns" have confronted the hazards of our time.

The agents of change in these stories also included the journalists and media companies that covered disasters and magnified their impact. All of these tragedies received extensive news coverage and public commentary that augmented their power as *focusing events*. In the age of steam-driven

engines, newspapers described workplace disasters in the most terrifying terms, telling readers just how badly bodies were burned, mangled, and torn asunder. In the era of television, when photojournalists made careers out of filming calamities, the news accounts were even more sensational and graphic. Now, the public actually *saw* disasters unfold. Burning buildings, dramatic rescue operations, and survivors' distraught faces captivated television audiences, which translated into higher ratings. News networks learned to compress disasters into a set of frightening images accompanied by emotional soundtracks and bold banners. In doing so, they heightened the sense of drama that disasters generated.[17]

Spectacles like the MGM fire provided powerful incentives for businesses to embrace new regulatory policies. Business leaders recognized the importance of protecting their customer bases, even if it meant costly changes. Whenever catastrophes scared away consumers, they threatened entire industries. Even highly profitable corporations could not afford to dismiss these events as freak accidents. When large airliners fell from the sky or posh resorts burned, champions of laissez-faire economics often became safety advocates. Working through employer organizations such as the Air Transport Association and Nevada Resort Association, corporate leaders generally supported stronger safety regulations. At the same time, however, they tried to shape these regulations to their own advantage. Like policy entrepreneurs and news reporters, they too became agents of change.

Still other individuals and groups played important roles in post-disaster policy making. In some cases, employees and their unions elevated safety issues. During congressional hearings that followed the Grand Canyon tragedy, airline pilot Jasper Solomon and his union's president, C. N. Sayen, challenged the perception that air travel was safe. Sayen even placed the blame for the air tragedy on lawmakers themselves. The testimony of Vern Balderston before Nevada legislators was equally powerful. Representing Nevada's firefighters, Balderston attributed all of the recent resort fire fatalities to the state's lax fire safety codes. Local communities and their spokesmen pushed for policy change in other ways. After the tragedy in Texas City, for example, more than four hundred local residents sued the federal government, claiming that its officials had been negligent in overseeing a postwar foreign aid program. To call attention to the problems faced after the 1971 earthquake, local residents in the San Fernando Valley held a mass rally. Meanwhile,

Reverend Francis Weber lamented the quake's devastating impact on the cherished San Fernando Mission and its treasures.

Policy-making debates proved enlightening in several ways. To begin with, they showed that new safety standards were more than by-products of singular, isolated events. All of these disasters had predecessors that could have been seen as warning signs. The devastation in Texas City stunned the nation, but the lethality of ammonium nitrate had been known since the early 1920s, when a massive explosion at a German chemical company on the Rhine River killed more than five hundred people and injured another two thousand.[18] Similarly, several midair collisions preceded the disaster over the Grand Canyon, and several previous earthquakes exposed the vulnerability of Southern California's hospitals to seismic activity. The tragedy at the MGM might have been predicted, too. There were more than a dozen deadly hotel fires in the United States during the 1970s, claiming a total of nearly two hundred lives. The Oklahoma City bombing also had its predecessors. In 1983, terrorists in Beirut, Lebanon, used truck bombs to attack the living facilities of a peacekeeping force that included some 300 US marines, 220 of whom lost their lives. Just as troubling were multiple threats made in 1994 to bomb the Internal Revenue Service, all of which reflected a growing anti-government sentiment within the nation.[19]

These signs convinced at least a few policymakers to begin the process of analyzing threats to public safety and to start developing ideas that were eventually brought forward as potential solutions to safety problems. Earthquakes during the 1960s, for example, persuaded a few policy protagonists to draw up lists of seismic safety recommendations, including proposals to map fault lines, strengthen building code requirements, and make gas and oil companies install automatic shutoff valves on their pipelines. By 1971, in the words of political scientists, these proposals were "on the shelf." Coalitions of technical experts and public officials had come together to address California's seismic safety problem. This was all part of the "primeval soup" from which policymakers drew to strengthen safety regulations. Unfortunately, it took more misery to overcome the inertia that had previously inhibited seismic safety reforms.[20]

In all of these cases, the reform process proceeded methodically, often all too slowly, even when parties agreed on the need for new safety regula-

tions. It took three years to overhaul the nation's air transportation system after the Grand Canyon crash, and even longer to create and implement special seismic safety standards for California hospitals after the 1971 earthquake. Nevada passed its comprehensive fire safety act within a few months of the MGM tragedy, but the federal government waited several more years before pressuring national hotel chains to adopt similar measures. In the wake of the Oklahoma City bombing, President Clinton directed the Justice Department to recommend new safety standards for all federal facilities, but the high-security structure that replaced the Murrah Building did not open for nearly a decade.

Real reform required coordination and interaction between countless numbers of individuals, groups, and organizations. Addressing the problem of uncontrolled airspace brought together professional associations of pilots, commercial airlines, air traffic controllers, and radio operators. It also required input from the armed forces and the support of congressional committees and their subcommittees. Strengthening seismic safety codes for hospitals in California depended on the work of seismologists, geologists, structural engineers, architects, hospital administrators, and several other groups. It also involved city councils, county commissioners, state senators, and the governor. Members of Congress and leaders of federal agencies got caught up in this movement, too, and turned it into a national endeavor. This was not extraordinary or strange but rather the nature of the modern policy-making process. Making progress took time. Everyone had to be consulted. Public approval mattered. The dedicated individuals and agencies that supported safety reforms were the unsung heroes in these stories. Like social activists who had railed against conditions in old factories and mines, they represented the public conscience.

The rules and regulations that arose from these disasters generally proved effective. In each case, reforms addressed basic underlying problems that the tragedies revealed. Those that emerged from the Texas City disaster tackled the careless ways in which the nation had manufactured, distributed, and stored ammonium nitrate, and they stopped ships from bringing such explosive cargo into the nation's harbors. The rise of the Federal Aviation Agency in the late 1950s brought an end to an old era in air traffic management, a time when airline pilots flew half blind over large parts of the country. New fire safety standards of the 1980s proved especially effective. Like Nevada, states recognized that local businesses had failed to develop sensible

fire safety systems and that the state itself had not established appropriate fire codes and fire safety boards. This framework of codes, laws, and agencies formed the proverbial silver lining in these stories. To be sure, each of these disasters spurred significant reforms that individually and collectively registered as progress.[21]

Safety reforms had their limits, of course. Time after time, new safety systems have only minimized or mitigated disasters; they rarely prevented them altogether. Even the newest, most technically complex devices and comprehensive regulations have failed to make work environments completely fireproof or otherwise immune to danger. An air disaster over New York City in 1960 illustrated the point. Despite the extensive air safety reforms of the late 1950s, two large commercial airliners collided in midair as they descended into LaGuardia Airport. The accident killed 140 people, including 8 on the ground.[22] The 1994 earthquake that rattled San Fernando Valley reminded Californians that even their best scientists and public leaders had not made homes and buildings completely quake resistant. The fact that the tremor did not wreck the valley's hospitals meant little to the people who lost loved ones in the tragedy.

Assessing the real value of such reforms invites a foray into counterfactual history in which we try to imagine tragedies that never occurred. No one can accurately count the number of lives that modern safety reforms have saved or the injuries they prevented. Nor can we easily estimate how much these reforms protected property values or levels of production. This problem makes it harder to appreciate the significance of previous safety reforms and to generate interest in future safety movements. Perhaps no one knew that better than Kofi Annan, the former secretary general of the United Nations. In his 1999 annual report on the organization's activities, Annan called attention to the challenge of promoting disaster-prevention campaigns. "Building a culture of prevention is not easy," the Ghanaian diplomat lamented. "While the costs of prevention have to be paid in the present, its benefits lie in a distant future. Moreover, the benefits are not tangible; they are the disasters that did not happen."[23] This is not to end on a cautionary note, but rather to remind readers that making work sites safer is a worthwhile and never-ending process. In a world seemingly driven by disasters (and now a global pandemic), safety tools and regulations always need to be maintained and, if necessary, improved.[24]

↓

The cultural critic Marshall Berman was certainly right when he said, "Grave danger is everywhere, and may strike at any moment." Berman's words speak to a central challenge of our time—dealing with the hazards of the modern world. As he explained all so well, the development of modern technologies and industries has created a smorgasbord of new opportunities while also generating an array of new hazards and risks. "To be modern," Berman wrote, "is to find ourselves in an environment that promises us adventure, power, joy, growth, transformation of ourselves and the world—and, at the same time, that threatens to destroy everything we have, everything we know, everything we are."[25] The word "modern" hints at something new and improved, bigger and better, faster and finer. It brings to mind sleek transit systems, towering skyscrapers, and other impressive structures. Symbols of advanced societies, however, need to be viewed from different angles. On closer scrutiny, tragic experiences appear to undergird all things modern, especially work environments.

If disasters are not the "mother" of modernity, they are surely the mother of safety reforms. In their wake, we can see the trails of two powerful, intimately related forces. Just as gravity keeps planets orbiting around stars, these forces keep us in a state of perpetual motion. Their winds have swept us up, tossed us about, and sometimes obliterated our workplaces. Yet they have also inspired us, brought us together, and often made us safer and more secure. To bring these forces into sharper focus, Berman pictured a strange "dialectical dance" in which two mysterious figures constantly transform modern landscapes. The apparition offers a fitting final scene: imagine yourself on a grand ballroom floor beneath large crystal chandeliers and gilded ceilings. Two dancers are off in the distance, twirling here and there, spinning side by side, dipping and turning your way. The scene is mesmerizing, even hypnotizing. But this is no genteel affair where gentlemen doff their hats to ladies. One of the dancers is a beast, destroying everything in its path. The other, a more ambiguous figure, is desperately clearing away spaces and creating the room anew. This mental picture—with its radically contradictory images—is deeply embedded in modern literature and art, and more of us might be imagined within its frame. After all, there are so many of these fateful dances.[26]

Notes

Introduction

1. James P. Kraft, *Vegas at Odds: Labor Conflict in a Leisure Economy, 1960–1985* (Baltimore: Johns Hopkins University Press, 2010); idem, "Disaster in the Workplace: The MGM Grand Hotel Fire of 1980," *Nevada State Historical Quarterly* 56, no. 3–4 (Fall–Winter 2013): 167–86.

2. For a sampling of the relevant historical literature, see Steven Biel, ed., *American Disasters* (New York: New York University Press, 2001); and Greg Bankoff, Uwe Lübken, and Jordan Sand, eds., *Flammable Cities: Urban Conflagration and the Making of the Modern World* (Madison: University of Wisconsin Press, 2012). For single-authored studies of multiple disasters, see Peter Charles Hoffer, *Seven Fires: The Urban Infernos That Reshaped America* (New York: Public Affairs, 2006); Jacob A. C. Remes, *Disaster Citizenship: Survivors, Solidarity, and Power in the Progressive Era* (Urbana: University of Illinois Press, 2016); and Kevin Rozario, *The Culture of Calamity: Disaster and the Making of Modern America* (Chicago: University of Chicago Press, 2007). On industry-wide approaches to the subject, see Mark Aldrich, *Death Rode the Rails: American Railroad Safety and Accidents, 1828–1965* (Baltimore: Johns Hopkins University Press, 2008); and Paul F. Paskoff, *Troubled Waters: Steamboat Disasters, River Improvements, and American Public Policy, 1821–1860* (Baton Rouge: Louisiana State University Press, 2007).

3. Ace Collins, *Tragedies of American History: 13 Stories of Human Error and Natural Disaster* (New York: Plume, 2003); Douglas Newton, ed., *Disaster, Disaster, Disaster: Catastrophes Which Changed Laws* (New York: Franklin Watts, 1961).

4. Henry George, *Progress and Poverty: An Inquiry into the Cause of Industrial Depressions and of Increase of Want with Increase of Wealth* (New York: D. Appleton, 1879), 8.

5. Henry George, *Social Problems* (New York: Henry George, 1886), 192–93.

6. On the field of risk theory, see Eugene A. Rosa, Ortwin Renn, and Aaron M. McCright, *The Risk Society Revisited: Social Theory and Governance* (Philadelphia: Temple University Press, 2014). See also Sheldon Krimsky and Dominic Golding, eds., *Social Theories of Risk* (London: Praeger, 1992), xii–xxiii, 1–9; and Arwen P. Mohun, *Risk: Negotiating Safety in American Society* (Baltimore: Johns Hopkins University Press, 2013), 7–8.

7. Ulrich Beck, *Risk Society: Towards a New Modernity* (London: Sage, 1992), 20, 100–101. (This is the English translation of Beck's book, which was originally published in 1986, in German.) On modernity and risks, see Anthony Giddens, *The Consequences of Modernity* (Stanford, CA: Stanford University Press, 1990). See also Ulrich Beck, Anthony Giddens, and Scott Lash, *Reflexive Modernization: Politics, Tradition and Aesthetics in the Modern Social Order* (Stanford, CA: Stanford University Press, 1994). On the role of government in managing risks, see David Moss, *When All Else Fails: Government as the Ultimate Risk Manager* (Cambridge, MA: Harvard University Press, 2002).

8. On definitions of the Southwest, see Raymond Mohl, ed., *Searching for the Sunbelt: Historical Perspectives on a Region* (Knoxville: University of Tennessee Press, 1990). See also Carl Abbott, *The Metropolitan Frontier: Cities in the Modern American West* (Tucson: University of Arizona Press, 1993), 157–59.

9. Gerald D. Nash, *The American West in the Twentieth Century: A Short History of an Urban Oasis* (Albuquerque: University of New Mexico Press, 1973), 191–213. Nash clarified his thesis in *The American West Transformed: The Impact of the Second World War* (Bloomington: Indiana University Press, 1985); *World War II and the West: Reshaping the Economy* (Lincoln: University of Nebraska Press, 1990); and *The Federal Landscape: An Economic History of the Twentieth-Century West* (Tucson: University of Arizona Press, 1999). See also Richard W. Etulain, *Beyond the Missouri: The Story of the American West* (Albuquerque: University of New Mexico Press, 2006), 362–448.

10. Nash, *American West in the Twentieth Century*, 193; idem, *American West Transformed*, 68–74. For more on postwar population growth, see also Abbott, *Metropolitan Frontier*, xvii–xxiii; and Richard M. Bernard and Bradley R. Rice, *Sunbelt Cities: Politics and Growth since World War II* (Austin: University of Texas Press, 1983). Also see Peter Wiley and Robert Gottlieb, *Empires in the Sun: The Rise of the New American West* (New York: G. P. Putnam's Sons, 1982). On early government-business relations in the West, see Mansel G. Blackford, *The Lost Dream: Businessmen and City Planning on the Pacific Coast, 1890–1920* (Columbus: Ohio State University Press, 1993).

11. On the 1944 naval munitions disaster and its aftermath, see Robert L. Allen, *The Port Chicago Mutiny* (New York: Warner Books, 1989); Leonard Guttridge, *Mutiny: A History of Naval Insurrection* (Annapolis: Naval Institute Press, 1992); and Charles Wollenberg, "Black vs. Navy Blue: the Mare Island Mutiny Court Martial," *California History* 58, no. 1 (Spring 1979): 62–75. See also *San Francisco Chronicle*, July 19, 1944.

12. Examples of home-front studies that explore the war's underside include Erasmo Gamboa, *Mexican Labor and World War II: Braceros in the Pacific Northwest, 1942–47* (Seattle: University of Washington Press, 1990); Roger Daniels, *Prisoners Without Trial: Japanese Americans in World War II* (New York: Hill and Wang, 1993); and Roger Daniels, Sandra C. Taylor, and Harry H. L. Kitano, eds., *Japanese Americans: From Relocation to Redress* (Salt Lake City: University of Utah Press, 1986). The 1944 zoot suit riots also mirrored this theme. Insightful works on the riots include Luis Alvarez, *The Power of the Zoot: Youth Culture and Resistance during World War II* (Berkeley: University of California Press, 2008); and Mauricio Mazon, *The Zoot Suit Riots: The Psychology of Symbolic Annihilation* (Austin: University of Texas Press, 1984). On labor-related issues, see William H. Harris, "Federal Intervention in Union Discrimination: FEPC and West Coast Shipyards during World War II," *Labor History* 22 (Summer 1981): 325–47. Another useful study is Arthur C. Verge, *Paradise Transformed: Los Angeles during the Second World War* (Dubuque, IA: Kendall/Hunt, 1995).

13. See Roger W. Lotchin, "California Cities and the Hurricane of Change: World War II in the San Francisco, Los Angeles, and San Diego Metropolitan Areas," *Pacific Historical Review* 63, no. 3 (August 1994): 393–420; "The Historians' War or the Home Front's War? Some Thoughts for Western Historians," *Western Historical Quarterly* 26 (Summer 1995): 185–96; "World War II and Urban California: City Planning and the Transformation Hypothesis," *Pacific Historical Review* 62, no. 2 (May 1993): 143–71. For more detail, see Lotchin's

Fortress California, 1911–1960: From Warfare to Welfare (New York: Oxford University Press, 1992); and *The Bad City in the Good War: San Francisco, Los Angeles, Oakland, and San Diego* (Bloomington: University of Indiana Press, 2003). See also James N. Gregory's evaluation of Nash's thesis in "Old West, New West: World War II and the Transformation of the American West," *Reviews in American History* 19, no. 2 (June 1991): 249–54.

14. This "booster class" was particularly successful in Phoenix. Between 1948 and 1964, more than seven hundred manufacturing firms moved into the greater Phoenix area, the bulk of which were in relatively new economic fields like aerospace engineering, light electronics, and high-tech research and development. See Elizabeth Tandy Shermer, *Sunbelt Capitalism: Phoenix and the Transformation of American Politics* (Philadelphia: University of Pennsylvania Press, 2013).

15. Shermer, *Sunbelt Capitalism*, 73–90, 122–30. In the war's early stages, for example, civic leaders had purchased and leased a large airfield to the army, where nearly 150,000 pilots received military training during the war years. At roughly the same time, private companies in the area began producing small planes, pontoon bridges, and other war-related commodities. By 1945, Phoenix was a center of defense production. For more on the rise of postwar Phoenix, see Andrew Needham, *Power Lines: Phoenix and the Making of the Modern Southwest* (Princeton, NJ: Princeton University Press, 2014).

16. For an overview of major works and themes in this field, see Michael K. Lindell, "Disaster Studies," *Sociopedia.isa*, 2011, https://sociopedia.isaportal.org/resources/resource /disaster-studies/.

17. See Charles E. Fritz, "Disaster," in *Contemporary Social Problems*, ed. Robert K. Merton and Robert A. Nisbet (New York: Harcourt, Brace and World, 1961), 655–56. See also Fritz's obituary in the *Washington Post*, May 7, 2000.

18. On definitions of disasters, see Henry W. Fisher, "The Sociology of Disaster: Definitions, Research Questions, and Measurements," *International Journal of Mass Emergencies and Disasters* 21, no. 1 (March 2003): 91–107; and Lindell, "Disaster Studies," 1–2.

19. When Fritz wrote that disasters were events "concentrated in time and space," he quoted the work of sociologist Robert Endleman. Concerning the life histories of disasters, Fritz spoke of periods of threat, impact, and post-impact. See Fritz, "Disaster," 655–82.

20. John W. Kingdon, *Agendas, Alternatives, and Public Policies*, 2nd ed. (New York: Longman, 1994), 21, 99–103, 177–78, 188–93. For an overview on this literature, see Thomas A. Birkland and Megan K. Warnement, "Focusing Events, Risk, and Regulation," *In Policy Shock: Recalibrating Risk and Regulation after Oil Spills, Nuclear Accidents and Financial Crises*, ed. Edward J. Balleisen et al. (New York: Cambridge University Press, 2017), 107–28. See also Debra Thompson and Jennifer Wallner, "A Focusing Tragedy: Public Policy and the Establishment of Afrocentric Education in Toronto," *Canadian Journal of Political Science* 44, no. 4 (December 2011): 807–28.

21. Thomas A. Birkland, *After Disaster: Agenda Setting, Public Policy, and Focusing Events* (Washington D.C.: Georgetown University Press, 1997), 18–19. See also idem, *An Introduction to the Policy Process: Theories, Concepts, and Models of Public Policy Making* (New York: M. E. Sharpe, 2001); *Lessons of Disaster: Policy Change after Catastrophic Events* (Washington, DC: Georgetown University Press, 2006).

22. Paul Sabatier and Hank C. Jenkins-Smith, "The Advocacy Coalition Framework: An

Assessment," in *Theories of the Policy Process*, ed. Paul Sabatier (Boulder, CO: Westview Press, 1999), 117–66. See also Daniel P. Carpenter, *Reputation and Power: Organizational Image and Pharmaceutical Regulation at the FDA* (Princeton, NJ: Princeton University Press, 2010).

23. Ingar Palmlund, "Social Drama and Risk Evaluation," in *Social Theories of Risk*, ed. Sheldon Krimsky and Dominic Golding (Westport, CT: Praeger, 1992), 197–212. On disasters as social dramas, see also Fritz, "Disaster," 691.

24. On technology and policy making, see Rosa et al., *Risk Society Revisited*, 78–79, 117, 124–25; Daniel A. Mazmanian and Paul A. Sabatier, *Implementation and Public Policy* (New York: University Press, 1989), 23, 31; and Nikolaos Zahariadis, "Ambiguity, Time, and Multiple Streams," in *Theories of the Policy Process*, ed. Paul Sabatier (Boulder, CO: Westview Press, 1999), 83.

25. Birkland, *After Disaster*, 14.

26. Joseph A. Schumpeter, *Capitalism, Socialism, and Democracy* (New York: Harper & Row, 1950), 84.

27. Schumpeter, *Capitalism, Socialism, and Democracy*, 32.

28. Schumpeter, *Capitalism, Socialism, and Democracy*, 84.

29. Mark R. Wilson, *Destructive Creation: American Business and the Winning of World War II* (Philadelphia: University of Pennsylvania Press, 2016).

30. Rozario, *Culture of Calamity*, 3, 8, 20, 84–85. Historical studies of environmental disasters have proved exceptionally influential. See Donald Worster, *Dust Bowl: The Southern Plains in the 1930s* (New York: Oxford University Press, 1979); William Cronon, ed., *Uncommon Ground: Rethinking the Human Place in Nature* (New York: W. W. Norton, 1995); and Douglas Brinkley, *The Great Deluge: Hurricane Katrina, New Orleans and the Mississippi Gulf Coast* (New York: HarperCollins, 2006).

31. Rozario, *Culture of Calamity*, 20. See also Rozario's "Rising from the Ruins," *Wall Street Journal*, January 16, 2010.

32. For a useful overview of modernity as a concept, see Matthew J. Lauzon, "Modernity," in *The Oxford Handbook of World History*, ed. Jerry H. Bentley (New York: Oxford University Press, 2011), 72–88. On the literature of modernism, see also Rozario, *Culture of Calamity*, 10–20; and Marshall Berman, *All That Is Solid Melts into Air: The Experience of Modernity* (New York: Penguin, 1988), 15–36.

33. The oft-heard expression "order out of chaos" has its roots in the Latin phrase *Ordo Ab Chao*, a motto apparently used by medieval fraternities of stonemasons. Nietzsche wrote, "One must still have chaos in oneself to be able to give birth to a dancing star." *Thus Spoke Zarathustra* (New York: Penguin, 1961), 46. See also Nietzsche's *Beyond Good and Evil: Prelude to a Philosophy of the Future*, trans. Marion Faber (Oxford: Oxford University Press, 1998), 6.

34. Robert Louis Stevenson, *Strange Case of Dr. Jekyll and Mr. Hyde* (New York: Oxford Classics, 2006). There have been numerous theatrical renderings of this diabolical tale.

35. Ignatius Donnelly, *Caesar's Column: A Story of the Twentieth Century* (Middleton, CT: Wesleyan University Press, 2003). Also Edward Bellamy, *Looking Backward, from 2000 to 1887* (Bedford, MA: Applewood Books, 1888); and Upton Sinclair, *The Jungle* (New York: Doubleday and Page, 1906).

36. Ignatius Donnelly, *Ragnarok: The Age of Fire and Gravel* (New York: D. Appleton, 1883), 439. For more insights into Donnelly and his writings, see Frederic Cople Jaher, *Doubters and Dissenters: Cataclysmic Thought in America, 1885–1918* (New York: Macmillan, 1964), 96–123. Other novels of this sort include H. G. Wells, *The World Set Free* (New York: E. P. Dutton, 1914), a story about a devastating global war that was followed by the rise of a new, peaceful world republic. Jack London's *The Iron Heel* (New York: Macmillan, 1908), had predicted the opposite scenario. On disasters in literature, see Jennifer Travis, *Danger and Vulnerability in Nineteenth-Century American Literature: Crash and Burn* (New York: Lexington Books, 2018).

37. These short films are easily found online. The theme of destruction and re-creation defined several other well-known silent films, such as Segundo de Chomón's *The Golden Beetle* (1906), Sergei Eisenstein's *Battleship Potemkin* (1925), and Fritz Lang's *Metropolis* (1927).

38. Berman, *All That Is Solid Melts into Air*, 295–96. For an overview of Berman's writings, see William Yardley's tribute to the philosopher, *New York Times*, September 15, 2013. See also Berman's *Modernism in the Streets: A Life and Times in Essays*, ed. David Marcus and Shellie Sclan (New York: Verso Books, 2017).

39. Berman, *All That Is Solid Melts into Air*, 15, 23, 48, 95, 101–3, 141, 288–91, 205–96, 313, 348. The construction of a massive expressway through his old neighborhood in the Bronx affected Berman on a personal level. As he noted, the construction displaced an entire community.

40. See Giddens, *Consequences of Modernity*, 7–10; and Beck, *Risk Society*, 19–27, 67.

Chapter 1. Disasters at Work

Epigraph. Ulrich Beck, *Risk Society: Towards a New Modernity* (London: Sage, 1992), 22.

1. On industrial accidents in England, see Jamie L. Bronstein, *Caught in the Machinery: Workplace Accidents and Injured Workers in Nineteenth-Century Britain* (Stanford, CA: Stanford University Press, 2008). See also Friedrich Engels, *The Condition of the Working Class in England* (Stanford, CA: Stanford University Press, 1968), 184–87, 282–84, 292; and P. E. H. Hair, "Mortality from Violence in British Coal-Mines, 1800–50," *Economic History Review* 21, no. 3 (December 1968): 545–61.

2. Walter Johnson described steamboat explosions as "the nineteenth century's first confrontation with industrialized mayhem." See *River of Dark Dreams: Slavery and Empire in the Cotton Kingdom* (Cambridge, MA: Belknap Press of Harvard University Press, 2013), 111. On the size and composition of steamboat workforces, see Louis C. Hunter's *Steamboats on the Western Waters: An Economic and Technological History* (Cambridge, MA: Harvard University Press, 1949), 446–50. An enduring work on steamboat disasters is James Lloyd's *Steamboat Directory and Disasters on the Western Waters* (Cincinnati: J. T. Lloyd, 1856).

3. Mark Twain and Charles Dudley Warner, *The Gilded Age* (New York: American Publishing Company, 1873).

4. On business risks and liability laws in this period, see Arwen P. Mohun, *Risk: Negotiating Safety in American Society* (Baltimore: Johns Hopkins University Press, 2013), 92–94; and David A. Moss, *When All Else Fails: Government as the Ultimate Risk Manager* (Cambridge, MA: Harvard University Press, 2002), 53–84.

5. Hunter, *Steamboats on the Western Waters*, 284–87; Lloyd, *Steamboat Directory and Disasters*, 89–93. Lloyd suggests that the *Moselle* tragedy killed eight of the crew and left five other crewmembers badly injured. See also Paul F. Paskoff, *Troubled Waters: Steamboat Disasters, River Improvements, and American Public Policy, 1821–1860* (Baton Rouge: Louisiana State University Press, 2007), 19–20.

6. John G. Burke, "Bursting Boilers and the Federal Power," *Technology and Culture* 7, no. 1 (1966): 1–23; Jerry L. Mashaw, *Creating the Administrative Constitution: The Lost One Hundred Years of American Administrative Law* (New Haven, CT: Yale University Press, 2012), 187–208.

7. John W. Kingdon, *Agendas, Alternatives, and Public Policies* (New York: HarperCollins, 1984), 21, 101.

8. Hunter, *Steamboats on the Western Waters*, 286–88. The referenced story concerned the ill-fated steamship the *Clipper*. The story was apparently reprinted in multiple papers, including Boston's *Advertiser*, October 4, 1843.

9. Hunter, *Steamboats on the Western Waters*, 288.

10. Mashaw, *Creating the Administrative Constitution*, 189–208.

11. Mark Aldrich, *Death Rode the Rails: American Railroad Accidents and Safety, 1828–1965* (Baltimore: Johns Hopkins University Press, 2006), 18–25. For more on the subject, see Aldrich's *Back on Track: American Railroad Accidents and Safety, 1956–2015* (Baltimore: Johns Hopkins University Press, 2018).

12. The train's engineer and firemen were among the victims. Charles Francis Adams, *Notes on Railroad Accidents* (New York: G. P. Putnam's Sons, 1879), 58–67.

13. Aldrich, *Death Rode the Rails*, 38–40; Adams, *Notes on Railroad Accidents*, 89–98.

14. Aldrich, *Death Rode the Rails*, 103–7. See also Steven W. Usselman, *Regulating Railroad Innovation: Business, Technology, and Politics in America, 1840–1920* (New York: Cambridge University Press, 2002), 304–5, 316–17.

15. Adams, *Notes on Railroad Accidents*, 138–40.

16. Adams, *Notes on Railroad Accidents*, 138–40. Bad weather conditions also explained the Revere tragedy. At the time of the crash, a light rain was blowing in from the sea, which made it difficult to see the taillights of the smaller train and harder for the larger train to stop. On the evolving relationship between rail accidents and methods of managing rail traffic, see Benjamin Sidney Michael Schwantes, *The Train and the Telegraph: A Revisionist History* (Baltimore: Johns Hopkins University Press, 2019).

17. Thomas K. McCraw, *Prophets of Regulation* (Cambridge, MA: Harvard University Press, 1984), 29. On Adams and the Revere tragedy, see also Usselman, *Regulating Railroad Innovation*, 120–25. Between 1871 and 1874 the number of locomotives and railcars in Massachusetts equipped with automatic braking systems jumped from nearly zero to more than one thousand, and the number of cars designed not to swing or crush on impact rose from a negligible number to roughly eight hundred. See Adams, *Notes on Railroad Accidents*, 43–53; Steven W. Usselman, "Air Brakes for Freight Trains: Technological Innovation in the American Railroad Industry, 1869–1900," *Business History Review* 58 (Spring 1984): 30–50.

18. Aldrich, *Death Rode the Rails*, 130–31, 143; and Adams, *Notes on Railroad Accidents*, 98–111.

19. Aldrich, *Death Rode the Rails*, 143–54.

20. Aldrich, *Death Rode the Rails*, 70–96, 111–15. On the 1893 Safety Appliance Act, see also Usselman, *Regulating Railroad Innovation*, 292.

21. These figures are available at "Achievements in Public Health, 1900–1999: Improvements in Workplace Safety—United States, 1900–1999," Centers for Disease Control and Prevention, accessed May 22, 2020, https://www.cdc.gov/mmwr/preview/mmwrhtml/mm 4822a1.htm.

22. Joseph P. Goldberg and William T. Moye, *The First Hundred Years of the Bureau of Labor Statistics* (Washington, DC: US Department of Labor, 1985), 58. On the rise of the National Safety Council, see Arwen Mohun, *Risk: Negotiating Safety in American Society* (Baltimore: Johns Hopkins University Press, 2013), 137–40. For more on the period's fatality rates and the law, see Moss, *When All Else Fails*, 162–69.

23. On conditions in mines, see Crystal Eastman, *Work-Accidents and the Law*, vol. 2 of *Pittsburgh Survey*, ed. Paul Underwood Kellogg (Philadelphia: Russell Sage Foundation, 1910), 34–48. For more on occupational injuries and fatalities in mining, see Thomas Dublin and Walter Licht, *The Face of Decline: The Pennsylvania Anthracite Region in the Twentieth Century* (Ithaca, NY: Cornell University Press, 2005); David Alan Corbin, *Life, Work, and Rebellion in the Coal Fields: The Southern West Virginia Miners, 1880–1922* (Urbana: University of Illinois Press, 1981).

24. An unknown number of boys were in the Monongah mines, and the death toll likely exceeded four hundred. See the verdict of the coroner's jury in "Monongah Mine Disaster," *Illustrated Monthly West Virginian* (January 1908): http://www.wvculture.org/history /disasters/monongah03.html.

25. Useful book-length studies of coal mine practices include Keith Dix, *What's a Coal Miner to Do? The Mechanization of Coal Mining* (Pittsburgh: University of Pittsburgh Press, 1988); Anthony F. C. Wallace, *St. Clair: A Nineteenth-Century Coal Town's Experience with a Disaster-Prone Industry* (New York: Alfred A. Knopf, 1987); James Whiteside, *Regulating Danger: The Struggle for Mine Safety in the Rocky Mountain Coal Industry* (Lincoln: University of Nebraska Press, 1990); Price Fishback, *Soft Coal, Hard Choices: The Economic Welfare of Bituminous Coal Miners, 1890–1930* (New York: Oxford University Press, 1992); and William Graebner, *Coal Mining Safety in the Progressive Period* (Lexington: University of Kentucky Press, 1976).

26. Eastman, *Work-Accidents and the Law*, 47–48.

27. See the Chicago newspapers for the week beginning November 14, 1909. Also, Karen Tintori, *Trapped: The 1909 Cherry Mine Disaster* (New York: Atria Books, 2002).

28. For a good look at mining in Montana, see Melvyn Dubofsky, *We Shall Be All: A History of the Industrial Workers of the World* (Chicago: Quadrangle Books, 1969), 301–7. The mine companies in Butte apparently printed "No Smoking" signs in sixteen different languages. Other useful studies include: James Whiteside, *Regulating Danger: The Struggle for Mine Safety in the Rocky Mountain Coal Industry* (Lincoln: University of Nebraska Press, 1990). For more context, see Dix, *What's a Coal Miner to Do?*; Wallace, *St. Clair*.

29. See Bill Haywood's *The Autobiography of William D. Haywood* (New York: International Publishers, 1929, 1958), 69, 80–81. See also Anthony Lukas, *Big Trouble: A Murder in a Small Western Town Sets Off a Struggle for the Soul of America* (New York: Touchstone, 1968).

Haywood is the central character in Dubofsky, *We Shall Be All*, and Dubofsky lists several works on the labor leader in his essay on sources (532).

30. Useful articles on safety problems in mining include Michael Wallace, "Dying for Coal: the Struggle for Health and Safety Conditions in American Coal Mining, 1930–82," *Social Forces* 66, no. 2 (December 1987): 336–64; Michael S. Lewis-Beck and John R. Alford, "Can Government Regulate Safety? The Coal Mine Example," *American Political Science Review* 74, no. 3 (September 1980): 745–56; Mark Aldrich, "Preventing 'The Needless Peril of the Coal Mine': The Bureau of Mines and the Campaign against Coal Mine Explosions, 1910–1940," *Technology and Culture* 36, no. 3 (1995): 483–518.

31. Congress did not pass the kind of legislation that safety advocates wanted until the late 1960s, after a grisly disaster in Farmington, West Virginia, killed another seventy-eight miners. The 1969 Coal Act, which President Lyndon B. Johnson had strongly supported, not only made mine safety standards more uniform from state to state but also gave federal agents the power to enforce those standards. After the law took effect, fatality rates in mining dropped sharply. By the 1980s, miners were nearly five times less likely to die on the job than their counterparts of the early twentieth century. For more on occupational safety in mining after World War II, see Bonnie E. Stewart, *No. 9: The 1968 Farmington Mine Disaster* (Morgantown: West Virginia University Press, 2011). On earlier postwar mining disasters, see Robert E. Hartley and David Kenney, *Death Underground: The Centralia and West Frankfort Mine Disasters* (Carbondale: Southern Illinois University Press, 2006).

32. On tragic events, media, and agenda setting, see Thomas A. Birkland, *After Disaster: Agenda Setting, Public Policy, and Focusing Events* (Washington, DC: Georgetown University Press, 1997), 1–14.

33. On labor militancy in western coal mining, see Thomas G. Andrews, *Killing for Coal: America's Deadliest Labor War* (Cambridge, MA: Harvard University Press, 2008). See also David A. Wolff, *Industrializing the Rockies: Growth, Competition, and Turmoil in the Coalfields of Colorado and Wyoming, 1868–1914* (Boulder: University Press of Colorado, 2003).

34. Eastman, *Work-Accidents and the Law*, 49–51. On smaller businesses in the industry, see John N. Ingham, *Making Iron and Steel: Independent Mills in Pittsburgh, 1820–1920* (Columbus: Ohio State University Press, 1991).

35. Eastman, *Work-Accidents and the Law*, 56–57. Eastman had worked closely with Florence Kelley, who had been appointed by Illinois Governor John Altgeld as his state's first chief factory inspector. On Kelley's pioneering role in the field of workplace safety, see Mohun, *Risk*, 126–32.

36. Eastman, *Work-Accidents and the Law*, 57–59, 70.

37. David Brody, *Steelworkers in America: The Nonunion Era* (Urbana: University of Illinois Press, 1998), 101–2, 167–69. As Brody explained, in times of sickness and health, steelworkers generally relied on assistance from branches of local groups such as the National Slovanic Society, the Polish National Alliance, and the Greek Catholic Union, to whom they paid monthly dues. See also Bronstein, *Caught in the Machinery*, 21–24; Mohun, *Risk*, 133–37; Moss, *When All Else Fails*, 152–69.

38. Eastman, *Work-Accidents and the Law*, 161–64. Many of the dependents of Carnegie's steelworkers lived in remote, rural parts of Europe, which made it difficult to prove they received support from the deceased.

39. On privately sponsored safety campaigns, see Mohun, *Risk*, 132–37.

40. On the Pemberton Mill, see Bronstein, *Caught in the Machinery*, 44–55, 66–68. Also, *New York Times*, January 11, 1860.

41. On the Grover Shoe Factory, see *New York Times*, March 21, March 30, and April 14, 1905. For a broader perspective on the shoe industry workers, see Mary H. Blewett, *Men, Women, and Work: Class, Gender, and Protest in the New England Shoe Industry, 1780–1910* (Urbana: University of Illinois Press, 1988); Alan Dawley, *Class and Community: The Industrial Revolution in Lynn* (Cambridge, MA: Harvard University Press, 1976).

42. Arthur F. McEvoy, "The Triangle Shirtwaist Factory Fire of 1911: Social Change, Industrial Accidents, and the Evolution of Common-Sense Causality," *Law and Social Inquiry* 20, no. 2 (Spring 1995): 621–51. Also, Richard Greenwald, *The Triangle Fire, the Protocols of Peace and Industrial Democracy in Progressive Era New York* (Philadelphia: Temple University Press, 2005); David Von Drehle, *Triangle: The Fire That Changed America* (New York: Grove Press, 2003). The Kheel Center at Cornell University created an outstanding website dedicated to the fire, "The 1911 Triangle Factory Fire," accessed May 22, 2020, https://triangle fire.ilr.cornell.edu/.

43. The owners survived the fire by running to the roof. On industrial accidents and laws in New York, see R. Rudy Higgens-Evenson, "From Industry Police to Workmen's Compensation: Public Policy and Industrial Accidents in New York, 1880–1910," *Labor History* 39, no. 4 (November 1998): 365–80. Also, Price Fishback and Shawn E. Kantor, *A Prelude to the Welfare State: The Origins of Workers' Compensation* (Chicago: University of Chicago Press, 2000); John Fabian Witt, "The Transformation of Work and the Law of Workplace Accidents, 1842–1910," Yale Law School, Faculty Scholarship Series, Paper 400, 1998, accessed June 13, 2020, https://digitalcommons.law.yale.edu/.

44. On perspectives and commemorations of the fire, see Paula E. Hyman, "Beyond Place and Ethnicity: The Uses of the Triangle Shirtwaist Fire," Jewish Women's Archives, accessed May 22, 2020, https://jwa.org/triangle/hyman.

45. Nat Brandt, *Chicago Death Trap: The Iroquois Theatre Fire of 1903* (Carbondale: Southern Illinois University Press, 2003), 3–28. A classic work on the fire is Marshall Everett, *The Great Chicago Theater Disaster* (Chicago: Publishers Union of America, 1904).

46. Brandt, *Chicago Death Trap*, 28–58. The floodlight that ignited the stage props and curtain was a carbon arc lamp with a cord.

47. Everett, *Great Chicago Theater Disaster*, 152–80.

48. Everett, *Great Chicago Theater Disaster*, 230–50. In St. Louis, Missouri, officials required theaters to add fire vents over stages that could be opened easily and instantly. In a few cases, local authorities arrested theater owners who failed to comply with new building codes. On the employment of theater musicians and stagehands in this period, see James P. Kraft, *Stage to Studio: Musicians and the Sound Revolution, 1890–1950* (Baltimore: Johns Hopkins University Press, 2003), 33–58.

49. See National Board of Fire Underwriters, *Report on the Cleveland Clinic Fire, Cleveland, Ohio, May 15, 1929* (New York: National Board of Fire Underwriters, 1929), https:// cplorg.contentdm.oclc.org/digital/collection/p128201collo/id/3512/.

50. National Board of Fire Underwriters, *Report on the Cleveland Clinic Fire*.

51. National Board of Fire Underwriters, *Report on the Cleveland Clinic Fire*.

52. On the school tragedy, see Ron Rozelle, *My Boys and Girls Are in There: The 1937 New London School Explosion* (College Station: Texas A&M Press, 2012); David M. Brown and Michael Wereschagin, *Gone at 3:17: The Untold Story of the Worst School Disaster in American History* (Dulles, VA: Potomac Books, 2012). Also, see Ace Collins, *Tragedies of American History: 13 Stories of Human Error and Natural Disaster* (New York: Plume, 2003), 37–52. The New London Museum maintains an informative website dedicated to the event: "The New London Texas School Explosion," accessed May 22, 2020, http://nlsd.net/index2.html.

53. On the Texas Engineering Practice Act, visit "Texas Board of Professional Engineers," Texas State Library and Archives Commission, accessed May 22, 2020, https://legacy.lib .utexas.edu/taro/tslac/50009/tsl-50009.html. See also "Licensing Information," Texas Board of Professional Engineers and Land Surveyors, accessed May 22, 2020, http://engineers .texas.gov/lic.htm.

54. On modern life as a two-sided phenomenon, see Anthony Giddens, *The Consequences of Modernity* (Stanford, CA: Stanford University Press, 1990), 7–10; Beck, *Risk Society*, 9–24. See also Marshall Berman, *All That Is Solid Melts into Air: The Experience of Modernity* (New York: Penguin, 1988), 15–36.

Chapter 2. Chemical Plant Explodes in Texas City!

Epigraph. Crystal Eastman, *Work-Accidents and the Law*, vol. 2 of *Pittsburgh Survey*, ed. Paul Underwood Kellogg (Philadelphia: Russell Sage Foundation, 1910), 13.

1. Significant books about the disaster include Hugh W. Stephens, *The Texas City Disaster 1947* (Austin: University of Texas Press, 1997); Bill Minutaglio, *City on Fire: The Forgotten Disaster That Devastated a Town and Ignited a Landmark Legal Battle* (New York: Harper-Collins, 2003); Tess Arrington, ed., *We Were There: A Collection of the Personal Stories of Survivors of the 1947 Ship Explosions in Texas City* (Texas City: Mainland Museum of Texas City, 1997). For an especially good overview of the disaster, see Milton MacKaye, "Death on the Waterfront," *Saturday Evening Post*, October 26, 1957, 19–21, 95–101.

2. For a brief summary of the disaster's significance, see Cathy Gillentine, "Lessons from the Disaster," *Texas City Sun*, April 13, 1997. Of the 568 people who lost their lives in the tragedy, only 405 were identified. The rest were classified as "unidentified dead" or "believed missing."

3. On the history of rubber, see John Tully, *Devil's Milk: A Social History of Rubber* (New York: Monthly Review Press, 2011).

4. American Chemical Society, "U.S. Synthetic Rubber Program," August 29, 1998, https:// www.acs.org/content/acs/en/education/whatischemistry/landmark/syntheticrubber /html.

5. American Chemical Society, "U.S. Synthetic Rubber Program."

6. On Germany's synthetic rubber program, see Tully, *Devil's Milk*, 283–317.

7. Chemical Heritage Foundation, "Rubber Matters: Solving the World War II Rubber Problem. Grain versus Petroleum Debate," accessed August 5, 2014, http://www.chemheritage .org/research/policy-center/oral-history-program/projects/rubber-matters/feature-grain -v-petroleum.aspx.

8. See "History of Monsanto," an unfinished study in Box 1, Folder 1.8, Texas City Di-

saster Records, Woodson Research Center Special Collection and Archives (hereafter cited as Woodson Research Center), Rice University, Houston, Texas.

9. "History of Monsanto."

10. "History of Monsanto." The tonnage of cargo that left Texas City's docks in the 1930s jumped from 3.5 million to 13.5 million. On the development of Texas City as a port, see Priscilla Benham, "Texas City: Port of Industrial Opportunity" (PhD diss., University of Houston, 1987), 43–66. The work is available at Moore Memorial Public Library, Texas City.

11. Benham, "Texas City," 219–39.

12. Benham, "Texas City," 219–39.

13. "Dedicatory Address of H. K. Eckert, April 8, 1949," vertical files, Folder: Texas City Disaster Monsanto Monument, Moore Memorial Public Library, Texas City.

14. "Dedicatory Address of H. K. Eckert."

15. "Dedicatory Address of H. K. Eckert."

16. "History of Monsanto."

17. "History of Monsanto." Monsanto purchased its Texas City plant from the federal government in August 1946 for approximately $9.5 million.

18. "History of Monsanto." Also, "Report by Fire Prevention and Engineering Bureau of Texas and the National Board of Fire Underwriters," 38–39, in vertical files, Folder: Texas City Disaster Reports, Moore Memorial Public Library, Texas City.

19. Monsanto Chemical Company Personnel Records (hereafter cited as Monsanto Personnel Records), Box 1, Folder 1.9, Texas City Disaster Records, Woodson Research Center, Rice University, Houston, Texas.

20. Federal Bureau of Investigation Records (hereafter cited as FBI Records), Box 43, Folder: Jacobs - Enestine Scholl, Record Group 118, Records of US Attorneys and Marshalls, Southern District of Texas, Houston Division (hereafter cited as Record Group 118), National Archives, Fort Worth, Texas.

21. FBI Records, Box 58, Folder: Merriam, Bernard, Record Group 118, National Archives, Fort Worth, Texas.

22. Monsanto Personnel Records, Box 1, Folder 1.9, Texas City Disaster Records, Woodson Research Center, Rice University, Houston, Texas.

23. FBI Records, Box 17, Folder: Dalrymple, Marie, Record Group 118, National Archives, Fort Worth, Texas.

24. Monsanto Personnel Records, Box 1, Folder 1.9, Texas City Disaster Records, Woodson Research Center, Rice University, Houston, Texas.

25. FBI Records, Box 33, Folder: Grice, Syble, Record Group 118, National Archives, Fort Worth, Texas.

26. Monsanto Personnel Records, Box 1, Folder 1.9, Texas City Disaster Records, Woodson Research Center, Rice University, Houston, Texas.

27. FBI Records, Box 31, Folder: Goff, Jim, Record Group 118, National Archives, Fort Worth, Texas.

28. Monsanto Personnel Records, Box 1, Folder 1.9, Texas City Disaster Records, Woodson Research Center, Rice University, Houston, Texas.

29. FBI Records, Box 3, Folder: Axe, Paul, Record Group 118, National Archives, Fort Worth, Texas.

30. "Report by Fire Prevention and Engineering Bureau of Texas," 38–39.

31. "Report by Fire Prevention and Engineering Bureau of Texas," 38–39.

32. FBI Records, Box 96, Folder: Yaos, Charles E., Record Group 118, National Archives, Fort Worth, Texas.

33. Arrington, *We Were There*, 41–42.

34. Arrington, *We Were There*, 21.

35. *Texas City Sun*, April 16, 2000.

36. FBI Records, Box 72, Folder: Robinson, J. D., Record Group 118, National Archives, Fort Worth, Texas.

37. Monsanto Personnel Records, Box 1, Folder 1.9, Texas City Disaster Records, Woodson Research Center, Rice University, Houston, Texas.

38. FBI Records, Box 3, Folder: Bailey, Robert, Record Group 118, National Archives, Fort Worth, Texas.

39. Most of the city's firefighters died on the docks when the *Grandcamp* exploded, leaving no one to battle the raging fires at oil refineries and the Monsanto plant. See John Ferling, "Texas City Disaster," *American History* 30, no. 6 (January/February 1996): 1–16.

40. *New York Times*, April 17, 1947.

41. Roy E. Doran to Asst Chief of Staff, April 16, 1947, vertical files, Folder: Texas City Disaster - Army Activities, Moore Memorial Public Library, Texas City.

42. J. C. Ford to Miss Mildred Stevenson, "Navy Assistance at Texas City," September 17, 1947, in vertical files, Folder: Texas City Disaster - Army Activities, Moore Memorial Public Library, Texas City. See also Leland R. Johnson, *Situation Desperate: U.S. Army Engineer Disaster Relief Operations, Origins to 1950* (Washington, DC: US Government Printing Office, 2011), 210–11.

43. "Brief Report of the Texas City Disaster," D. W. Griffiths, April 24, 1947, in vertical files, Folder: Texas City Disaster - Army Activities, Moore Memorial Public Library, Texas City.

44. Ferling, "Texas City Disaster," 9; *New York Times*, April 17, 1947.

45. Johnson, *Situation Desperate*, 209.

46. Edgar M. Queeny to Monsanto Stockholders, Employees and Friends, April 30, 1947, Box 1, Folder 1.2, Texas City Disaster Records, Woodson Research Center, Rice University, Houston, Texas. The plane's pilot later reported that black clouds from Texas City stretched all the way to Missouri.

47. Edgar M. Queeny to Monsanto Stockholders.

48. Edgar M. Queeny to Monsanto Stockholders.

49. The company's public relations department apparently stopped a considerable amount of misinformation from being distributed. See J. Hanley Wright to Executive Committee, "Report of the Department of Industrial and Public Relations," May 12, 1947, Box 1, Folder 1.4, Texas City Disaster Records, Woodson Research Center, Rice University, Houston, Texas.

50. Edgar M. Queeny to Monsanto Stockholders.

51. Edgar M. Queeny to Monsanto Stockholders.

52. Edgar M. Queeny to Monsanto Stockholders. The company established a working group to help displaced employees find temporary housing and to make repairs to the damaged homes of those workers. Monsanto's employees in other cities made their own donations to the disaster's victims, which proved significant. Employees at Monsanto's chemical plant in Oak Ridge, Tennessee, for example, contributed a day's pay to their Texas City counterparts, and other large donations came from the company's laboratories and sales offices in Alabama, California, Michigan, Missouri, Ohio, and other states. See also *New York Times*, April 17, 1947.

53. Edgar M. Queeny to Monsanto Stockholders.

54. "Monthly Report to Executive Committee From Texas Division, May 1947," Box 1, Folder 1.3, Woodson Research Center, Rice University, Houston, Texas.

55. FBI Records, Box 39, Folder: Holloway, Hannah, Record Group 118, National Archives, Fort Worth, Texas.

56. See FBI Records, Box 51, Folder: Lewis, John; Box 22, Folder: Earls, Elijah; and Box 18, Folder: Davis, James, Record Group 118, National Archives, Fort Worth, Texas.

57. FBI Records, Box 93, Folder: Weydert, Gussie, Record Group 118, National Archives, Fort Worth, Texas.

58. Ferling, "Texas City Disaster," 12.

59. Gillentine, "Lessons from the Disaster."

60. Gillentine, "Lessons from the Disaster." Also, "Report by Fire Prevention and Engineering Bureau of Texas," 38–39.

61. For a good overview of the 1950 act, see Kevin Rozario, *The Culture of Calamity: Disaster and the Making of Modern America* (Chicago: University of Chicago Press, 2007), 150–55.

62. 80 Cong. Rec. 7719 (1947).

63. As Senator Aiken told the Senate: "I have discussed the bill with the leaders of the Senate on both sides. I find no objection to it." See 80 Cong. Rec. 8300–8301 (1947). See also Johnson, *Situation Desperate*, 213–15.

64. See remarks of William Whittington on August 7, 1950, in 81 Cong. Rec. 11897 (1950).

65. Many plaintiffs combined their cases into *Elizabeth Dalehite et al. vs. United States* (1950). About fourteen hundred claims were ultimately approved, or more than 80 percent of those filed. See Texas City Disaster, H.R. Rep. No. 1386, at 111 (March 24, 1954).

66. The disaster also prompted the rise of a new full-service hospital in Galveston. On the role of surgeons in the disaster, see H. C. Slocum and C. R. Allen, "Anesthesia Services for the Texas City Disaster Patients," *Southern Medical Journal* 41, no. 4 (April 1948): 344–46, Truman G. Blocker Jr. History of Medicine Collections, Moody Medical Library, Galveston, Texas.

67. J. H. Hendrix, T. G. Blocker, and Virginia Blocker, "Plastic Surgery Problems in the Texas City Disaster," *Plastic and Reconstructive Surgery* 4, no. 1 (January, 1949): 92–97, in Truman G. Blocker Jr. History of Medicine Collections, Moody Medical Library, Galveston, Texas.

68. Hendrix et al., "Plastic Surgery Problems in the Texas City Disaster."

69. Low-income groups often settle in residential areas near enterprises like chemical

plants and oil refineries, where housing costs are relatively low. On class and risk distribution, see Ulrich Beck, *Risk Society: Towards a New Modernity* (London: Sage, 1992), 35–36, 52–53.

Chapter 3. Airliners Collide over the Grand Canyon!

Epigraph. Henry George, *Social Problems* (New York: Henry George, 1886), 12.

1. "Accident Investigation Report," Civil Aeronautics Board, April 17, 1957, 15–16, TWA Records, Collection Number KC-453, Folder 9, Box 145, State Historical Society of Missouri Research Center, University of Missouri–Kansas City, University Archives (hereafter cited as TWA Collection, UMKC Archives).

2. One of the best and most thorough studies of this tragedy is Mike Nelson's *We Are Going In: The Story of the 1956 Grand Canyon Midair Collision* (Tucson, AZ: Rio Nueva, 2012). See also Jon Proctor, "Midair! Lessons from a Tragedy over the Grand Canyon Fifty Years Later," *Wings* (August 2006): 17–26.

3. For biographical sketches of the pilots and their crews, see John Collings to All TWA Employees, July 2, 1956, KC-453, Folder 14, Box 145, TWA Collection, UMKC Archives.

4. In the words of John W. Kingdom, "focusing events need accompaniment." *Agendas, Alternatives, and Public Policies*, 2nd ed. (New York: Longman, 1994), 103.

5. On risk as a "traveling companion," see Eugene A. Rosa, Ortwin Renn, and Aaron M. McCright, *The Risk Society Revisited: Social Theory and Governance* (Philadelphia: Temple University Press, 2014), xxv–xxviii. On policy windows, see Kingdom, *Agendas, Alternatives, and Public Policies*, 174–79.

6. Gerald D. Nash, *The American West Transformed: The Impact of the Second World War* (Bloomington: Indiana University Press, 1985); Roger W. Lotchin, "California Cities and the Hurricane of Change: World War II in the San Francisco, Los Angeles, and San Diego Metropolitan Areas," *Pacific Historical Review* 63, no. 3 (August 1994): 393–420.

7. Marion F. Sturkey, *Mid-Air: Accident Reports and Voice Transcripts from Military and Airline Mid-Air Collisions* (Plum Branch, SC: Heritage Press International, 2008), 9–10. The army pilot bailed out of his own plane with a parachute, but the parachute got tangled in the wreckage of the plane.

8. There were approximately sixty-five midair collisions between 1950 and 1956. William B. Harding, "To Promote Safety in the Skies," *Air Line Pilot* (October 1956): 10.

9. Aviation Facilities Group, *Aviation Facilities: The Report to the Director, Bureau of the Budget* (Washington, DC: Bureau of the Budget, December 31, 1955), 15.

10. *Washington Post*, November 2, 1949. The pilot of the military plane, a Lockheed P-38 Lightning Fighter, reported engine trouble just before the collision. See also Sturkey, *Mid-Air*, 45–52.

11. *New York Times*, January 13, 1955. As investigators soon learned, Cincinnati's airport controllers had not cleared the smaller plane to enter the surrounding airspace or even made contact with the plane's pilot.

12. *Aviation Facilities*, 18.

13. Wolfgang Langewiesche, "Traffic Jam in the Sky," *Reader's Digest* (July 1956): 49–61. Also, Robert A. Stone, "Mid-Air Collisions: A Pilot's View," *Air Line Pilot* (February 1956): 3–5.

14. Lyle R. Hincks to Bart Hewitt, April 3, 1956, KC-453, Folder 10, Box 145, TWA Collection, UMKC Archives.

15. Lyle R. Hincks to Bart Hewitt.

16. Gerald D. Nash, *World War II and the American West: Reshaping the Economy* (Lincoln: University of Nebraska Press, 1990), 67–90.

17. See Harding, "To Promote Safety in the Skies." See also 102 Cong. Rec. 11897 (1956).

18. Nash, *World War II and the American West*, 67–70.

19. *Los Angeles Times*, November 19, 1995. Pan American Airlines also flew out of Los Angeles in the 1950s. See also Nathan Masters, "From Mines Field to LAX: The Early History of L.A. International Airport," July 25, 2012, https://www.kcet.org/shows/lost-la/from-mines-field-to-lax-the-early-history-of-la-international-airport.

20. Rudi Volti, *Technology and Commercial Air Travel* (Washington, DC: American Historical Association, 2015), 14–22. The CAB was also part of the Department of Commerce, though it acted much like an independent agency.

21. Volti, *Technology and Commercial Air Travel*, 77–78. Prior to CAB, regulatory power had rested in the Bureau of Air Commerce (BAC).

22. *Aviation Facilities*, 21. See also Staff, "RTCA: What It Is, What It Does," *Air Line Pilot* (December 1954): 11.

23. "CAA Fourth Region Statement Concerning the TWA-UAL Accident," July 2, 1956, KC-452, Folder 2, Box 144, TWA Collection, UMKC Archives. By 1956, the CAA had divided the nation's airspace into twenty-six areas, each of which was under the jurisdiction of a major metropolitan airport and its controllers. See also Staff, "Air Traffic Control—Tragedy in the Making," *Air Line Pilot* (August 1951): 2, 14.

24. Sturkey, *Mid-Air*, 64.

25. *Aviation Facilities*, 15–18.

26. Anchard F. Zeller, "Human Aspects of Mid-Air Collision Prevention," *Air Line Pilot* (December 1959): 4–9.

27. Staff, "ALPA Pledges Support of Two Air Safety Bills," *Air Line Pilot* (February 1950): 3, 7, 9. See also T. G. Linnert, "ALPA's Air Safety Development," *Air Line Pilot* (January 1956): 7–8, 23. Pilots organized ALPA in 1931. See "Constitution and By-Laws of the Air Line Pilots Association," KC-1383, Folder 15, Box 3, 3–5, TWA Active Retired Pilots Association (TARPA) Collection, UMKC Archives.

28. F. B. Lee, "Today's Air Navigation," *Air Line Pilot* (November 1955): 15–18.

29. Lee, "Today's Air Navigation."

30. Ed Modes, "Airborne Radar—A Step Forward," *Air Line Pilot* (November 1954): 4.

31. Modes, "Airborne Radar."

32. Hughes to Harding, May 4, 1955, in *Aviation Facilities*, iv–v.

33. 102 Cong. Rec. 11898 (1956).

34. 102 Cong. Rec. 11898 (1956). See also Harding, "To Promote Safety in the Skies," 10–12.

35. On Curtis, see *Time*, July 30, 1956.

36. For an in-depth look at the collision and its causes, see the CAA's Accident Investigation Report, KC-453, Folder 9, Box 145, TWA Collection, UMKC Archives. See also "Recorded Transmissions by Los Angeles Ground Control," KC-453, Folder 14, Box 145, TWA Collection, UMKC Archives.

37. "CAA Incident Report," July 1, 1956, KC-453, Folder 14, Box 145, TWA Collection, UMKC Archives. In the same box and folder, see also "Radio Logs for Flight 2," June 30, 1956.

38. "CAA Incident Report," July 1, 1956. Also, "TWA to Major Department Heads," July 2, 1956, KC-453, Folder 2, Box 144, TWA Collection, UMKC Archives.

39. *Airspace Use Study: Hearings before a Subcommittee of the Committee on Interstate and Foreign Commerce House of Representatives*, 84 Cong. 78–79 (July 18, 1956) (statement of Captain Lionel Stephan).

40. "TWA to Major Department Heads," July 2, 1956, KC-453, Folder 2, Box 144, TWA Collection, UMKC Archives. For more details on initial reaction to the news that the planes were missing, see Nelson, *We Are Going In*, 196–200.

41. "TWA to Major Department Heads." Also, *Deseret News*, July 2, 1956. To report the accident, the Hudgins drove fifteen miles to a pay phone in Grand Canyon Village. On the Hudgins, see Nelson, *We Are Going In*, 202–3, 228–29.

42. See reprints of articles about the crash site published in the *Arizona Republic*, April 5 and 6, 1977, available in KC-506, Folder 4, Gordon R. Parkinson Papers, 1929–1988, UMKC Archives.

43. John A. Collings to Townsend A. Peck, July 19, 1956, KC-453, Folder 3, Box 144, TWA Collection, UMKC Archives.

44. John C. Collings to Honorable Wilber M. Brucker, July 20, 1956, K-453, Folder 3, Box 144, TWA Collection, UMKC Archives. Brucker was secretary of the army.

45. An army officer who helped transport bodies to the burial site recalled the experience in a letter to Gordon "Parky" Parkinson, a TWA employee at the time of the crash and later president of the TWA Seniors Club. See Rod to Parky, February 2, 1978, KC-506, Folder 4, Gordon R. Parkinson Papers, 1929–1988, UMKC Archives.

46. Rod to Parky, February 2, 1978. About 350 people attended the burial service, held on July 9, 1956.

47. 102 Cong. Rec. 11897 (1956).

48. *Airspace Use Study*, 38.

49. *Airspace Use Study*, 43.

50. *Airspace Use Study*, 59–61.

51. *Airspace Use Study*, 62–63.

52. *Airspace Use Study*, 63, 67.

53. *Airspace Use Study*, 52, 93–98.

54. *Airspace Use Study*, 320–23.

55. *Airspace Use Study*. Congress had recently denied an industry request to add lights at the end of long runways.

56. *Airspace Use Study*, 304–5.

57. *Airspace Use Study*, 293–305. Furnas said the new technology would create a "common system," that is, one that would be acceptable to both the military and the commercial aviation business.

58. *Airspace Use Study*, 211–21. "The military, very credibly, has gone to a higher frequency band to relieve the congestion in the very high frequency radio band, and thus they get better communications," Pyle told the committee.

59. *Airspace Use Study*, 13. Flight analyzers contained no electronic circuitry. They could continue to operate ten minutes after a power failure and withstand exposure to two thousand degrees Fahrenheit for half an hour. See also 102 Cong. Rec. 11693 (1956).

60. *Airspace Use Study*, 2.

61. 103 Cong. Rec. 5500–5501 (1957).

62. 103 Cong. Rec. 5500–5501 (1957).

63. See E. R. Quesada to Oren Harris, July 17, 1957, 103 Cong. Rec. 13177 (1957).

64. 103 Cong. Rec. 13049 (1957).

65. 103 Cong. Rec. 13051 (1957).

66. 103 Cong. Rec. 13046–13048 (1957).

67. 103 Cong. Rec. 13046–13048 (1957).

68. See Garrison Norton, Assistant Secretary of the Navy, to Senator Thomas H. Kuchel, 104 Cong. Rec. 9197 (1958).

69. Sturkey, *Mid-Air*, 87–93.

70. Sturkey, *Mid-Air*, 94–98.

71. 104 Cong. Rec. 13645 (1958). Washington attorneys Robert Murphy and Thomas Finney helped Monroney's subcommittee write the bill.

72. The act gave the CAA the power to decide whether FAA rules imposed undue burdens on airline companies and other parties. The CAB remained responsible for investigating air accidents and regulating economic policies of commercial airlines. The act obligated the FAA to work with military advisors in regulating military flights. See provisions of the Federal Aviation Act of 1958, 104 Cong. Rec. 16861–16885 (1958).

73. See Message from the President of the United States, 104 Cong. Rec. 11149–11150 (1958).

74. 104 Cong. Rec. 13662 (1958). The amendments included limitations on the FAA's employment of consultants and language to ensure civilian control of the agency.

75. 104 Cong. Rec. 16087 (1958).

76. *Time*, February 22, 1960, 16–19.

77. Despite bitter opposition from the pilots' union, they had also adopted rules that required airline pilots to retire at age sixty, on the grounds that the potential for heart attacks and strokes among older pilots threatened public safety. See Staff, "FAA Sets Mandatory Age 60 Limit for Retirement of Airline Pilots," *Air Line Pilot* (January 1960): 24. Also, "Elwood 'Pete' Quesada: The Right Man for the Right Job," accessed May 26, 2020, www.faa .gov/about/history/heritage/media/Elwood_Quesada.pdf.

78. Kingdon, *Agendas, Alternatives, and Public Policies*, 101.

Chapter 4. Hospitals Collapse in Southern California!

Epigraph. Voltaire, "Poem on the Lisbon Disaster, 1756," in Joseph McCabe, trans., *Toleration and Other Essays by Voltaire* (New York: G. P. Putnam's Sons, 1912), 255–63.

1. A more powerful earthquake hit Santa Barbara County in 1977 but caused no loss of life and little property damage. On the early history of seismic activity in California, see US Department of Commerce, *Earthquake History of the United States, Part II* (Washington, DC: Government Printing Office, 1951), Box 319, Folder 1, Kenneth Hahn Collection, Huntington Library, Pasadena California (hereafter cited as Hahn Collection, Huntington Library).

On earthquakes and politics, see Carl-Henry Geschwind, *California Earthquakes: Science, Risk, and the Politics of Hazard Mitigation* (Baltimore: Johns Hopkins University Press, 2001). On Los Angeles and its vulnerability to disasters, see Mike Davis, *Ecology of Fear: Los Angeles and the Imagination of Disaster* (New York: Metropolitan Books, 1998).

2. For a better understanding of the relationship between politics, economics, and natural disasters, see Theodore Steinberg, *Acts of God: The Unnatural History of Natural Disaster in America* (Cambridge: Oxford University Press, 2000). See also Ari Kelman, "Nature Bats Last: Some Recent Works on Technology and Urban Disaster," *Technology and Culture* 47 (April 2006): 391–402. On how human activities set the stage for environmental disasters, see Donald Worster, *Dust Bowl: The Southern Plains in the 1930s* (New York: Oxford University Press, 1979).

3. Jean-Jacques Rousseau, "Letter to Voltaire," August 18, 1756, in *The Collected Writings of Rousseau*, ed. Roger D. Masters and Christopher Kelly (Hanover, NH: University Press of New England, 1992), 110. For a good translation of Voltaire's "Poem of the Lisbon Disaster," see Joseph McCabe, *Toleration and Other Essays by Voltaire* (New York: G. P. Putnam's Sons, 1912), 255–63. For an overview of this intellectual debate, see Russell R. Dynes, "The Dialogue between Voltaire and Rousseau on the Lisbon Earthquake: The Emergence of a Social Science View," *International Journal of Mass Emergencies and Disasters* 18, no. 1 (March 2000): 97–115. See also Kevin Rozario, *The Culture of Calamity: Disaster and the Making of Modern America* (Chicago: University of Chicago Press, 2007), 14–20.

4. Classic works on the subject of hospital labor include Barbara Melosh, *"The Physician's Hand": Work Culture and Conflict in American Nursing* (Philadelphia: Temple University Press, 1982); Leon Fink and Brian Greenberg, *Upheaval in the Quit Zone: A History of the Hospital Workers Union, Local 1199* (Urbana: University of Illinois Press, 1989); Karen Brodkin Sacks, *Caring by the Hour: Women, Work, and the Duke Medical Center* (Urbana: University of Illinois Press, 1998); and Jane E. Shultz, *Women at the Front: Hospital Workers in Civil War America* (Chapel Hill: University of North Carolina Press, 2004). The latter is one of several interesting studies on medical care during the Civil War.

5. County officials accepted title of the land in subdued ceremonies at the site. See *Valley News*, February 9, 1972.

6. On theater metaphors and policy making, see Ingar Palmlund, "Social Drama and Risk Evaluation," in *Social Theories of Risk*, ed. Sheldon Krimsky and Dominic Golding (Westport, CT: Praeger, 1992), 197–212; in the same collection of essays, see Ortwin Renn, "The Social Arena Concept of Risk Debates," 178–96.

7. On defense spending and population growth in California, see Carl Abbott, *The Metropolitan Frontier: Cities in the Modern American West* (Tucson: University of Arizona Press, 1993), 8–11, 57–63. Also, Gerald D. Nash, *The American West in the Twentieth Century: A Short History of an Urban Oasis* (Albuquerque: University of New Mexico Press, 1973), 191–95; idem, *World War II and the West: Reshaping the Economy* (Lincoln: University of Nebraska Press, 1990), 42, 68–90, 124–27, 131–34.

8. On the population of postwar cities in the West, see Nash, *American West in the Twentieth Century*, 213–15; Abbott, *Metropolitan Frontier*, 53–78.

9. David L. Clark, "Los Angeles: Improbable Los Angeles," in *Sunbelt Cities: Politics and*

Growth since World War II, ed. Richard M. Bernard and Bradley R. Rice (Austin: University of Texas Press, 1983), 287–90.

10. Clark, "Los Angeles."

11. *California Earthquake Disaster: Hearing before a Special Subcommittee of the Committee on Public Works House of Representatives* (hereafter cited as *California Earthquake Hearing*), 92 Cong. 46–47 (February 24, 1971) (statement of Robert J. Williams, director of building and safety, City of Los Angeles, 46–47).

12. *California Earthquake Hearing* (statement of Robert L. Prescott, coordinator, Housing of Urban Development Disaster Office, Los Angeles, 234).

13. *California Earthquake Hearing* (statement of Don Pool, manager, San Fernando Valley Human Resources Development Center, 239).

14. *California Earthquake Hearing* (statement of Harry Saunders, director, School Planning and Construction, Los Angeles Unified School District, 58). Also, *Van Nuys News*, February 11 and February 18, 1971.

15. Rev. Francis J. Weber, *Earthquake Memoir* (Hong Kong: Cathay Press, 1971).

16. The 1906 earthquake convinced San Francisco to approve the use of reinforced concrete in the building trades, which local brickmakers and their union had long opposed. Andrea Rees Davies, "The Social Impact of the 1906 San Francisco Earthquake and Fire," in *Flammable Cities: Urban Conflagration and the Making of the Modern World*, ed. Greg Bankoff, Uwe Lubken, and Jordon Sand (Madison: University of Wisconsin Press, 2012), 273–92.

17. *Earthquake History of the United States, Part II*, 27, Box 319, Folder 1, Hahn Collection, Huntington Library.

18. *Earthquake History of the United States, Part II*, 29. See also Steinberg, *Acts of God*, 38–39; Geschwind, *California Earthquakes*, 107.

19. Fred Turner, "Seventy Years of the Riley Act and Its Effect on California's Building Stock" (presented at the World Conference on Earthquake Engineers, Vancouver, BC, Canada, August 1–6, 2004), https://www.iitk.ac.in/nicee/wcee/article/13_313.pdf. "The great amount of damage that resulted from this shock was due to faulty construction of buildings," the coroner concluded. For more on the coroner's report, see Geschwind, *California Earthquakes*, 108.

20. The Field Act drew on recent legislation passed after the collapse of St. Francis Dam north of San Fernando Valley, which killed more than four hundred people. State of California, *Alfred E. Alquist, Seismic Safety Commission, The Field Act and Public School Construction: A 2007 Perspective* (Sacramento: State of California, February 2007), http://ssc.ca.gov/forms_pubs/cssc_2007-03_field_act_report.pdf. See also Robert A. Olson, "Legislative Politics and Seismic Safety: California's Early Years and the Field Act, 1925–1933," *Earthquake Spectra* 19, no. 1 (February 2003): 111–31, http://mitigation.eeri.org/files/Olson_Legislative_Politics_CA.pdf.

21. Turner, "Seventy Years of the Riley Act."

22. Turner, "Seventy Years of the Riley Act." For more on the Field and Riley Acts, see Geschwind, *California Earthquakes*, 113–14.

23. *Earthquake History of the United States, Part II*. Also, *Los Angeles Times*, June 2, 2002.

24. On the response to the Alaska earthquake, see Geschwind, *California Earthquakes*, 140–44.

25. *Los Angeles Times*, November 10, 1968.

26. See "Provisional Inventory, Biographical Information," Hahn Collection, Huntington Library.

27. Motion by Supervisor Kenneth Hahn, November 19, 1968, Box 319, Folder: Earthquakes—1963–64, 1968–69, Hahn Collection, Huntington Library.

28. *Summary of Earthquake Research Affiliates Conference*, May 18, 1967, Box 319, Folder: Earthquakes—1963–64, 1968–69, Hahn Collection, Huntington Library. Richter developed the method with the help of Beno Gutenberg, who also worked at the university. In short, the men developed an instrument known as a seismograph, which allowed anchored paper to unroll during a tremor. A pendulum suspended with a marking pen above the paper recorded the motion of earth. For more on Richter, see Geschwind, *California Earthquakes*, 98–99, 146–47, 185–87.

29. John Lambie to Board of Supervisors, September 4, 1969, Box 319, Folder: Earthquakes—1963–64, 1968–69, Hahn Collection, Huntington Library.

30. Karl Steinbrugge, Eugene Schader, and Donald Maran, "Building Damage in San Fernando Valley," in *San Fernando, California, Earthquake of 9 February 1971*, ed. Gordon B. Oakeshott, Bulletin 196 (Sacramento: California Division of Mines and Geology, 1975), 123–24. Older two-story dwellings fared much worse than their newer, single-level counterparts.

31. *Van Nuys News*, February 11, 1971. At the time the unemployment rate in California was already high. In San Fernando Valley, total number of unemployed topped two thousand.

32. D. F. Moran, "Damage to Energy and Communication Systems," in *San Fernando, California, Earthquake*, 245–48. The Departments of the Interior and Commerce produced this study. Also, *Los Angeles Times*, February 10, 1971.

33. *Los Angeles Times*, February 11, 1971, and February 4, 1996.

34. J. F. Meehan, "Damage to Transportation Systems," in *San Fernando, California, Earthquake*, 241. Also, *Los Angeles Times*, February 10, 1971.

35. *Los Angeles Times*, February 13, 1971, and February 4, 1996.

36. Steinbrugge et al., "Building Damage in San Fernando Valley," 349.

37. *Los Angeles Times*, February 10, 1971.

38. J. F. Heavey to Frank R. Hood, June 14, 1971, Box 2, Folder 2: San Fernando VA Hospital Earthquake February 9, 1971, Papers of Donald E. Johnson (hereafter cited as Donald Johnson Papers), Subject File, 1969–72, Richard Nixon Presidential Library and Museum, Yorba Linda, CA (hereafter cited as Nixon Library).

39. J. F. Heavey to Frank R. Hood, June 14, 1971.

40. J. F. Heavey to Frank R. Hood, June 14, 1971.

41. J. F. Heavey to Frank R. Hood, June 14, 1971.

42. *Los Angeles Times*, February 14, 1971. "We might have saved more lives if only we could have gotten to the disaster area sooner," the officer told the press.

43. *California Earthquake Hearing* (statement of Raymond M. Hill, chief engineer, Los Angeles City Fire Department, 53).

44. *California Earthquake Hearing* (statement of Raymond M. Hill, 53). See also J. F. Heavey to Frank R. Hood, June 14, 1971.

45. Colonel Robert J. Malley, April 28, 1971, Box 319, Folder: Earthquakes—1970–71, Hahn Collection, Huntington Library.

46. Colonel Robert J. Malley, April 28, 1971.

47. *Van Nuys News*, February 12, 1971; *San Francisco Chronicle*, February 13, 1971. See also Ted Thackrey Jr., "Survival a Kind of Miracle," *Los Angeles Times*, February 12, 1971.

48. *Van Nuys News*, February 12, 1971; *San Francisco Chronicle*, February 13, 1971.

49. "In Memoriam," San Fernando Veteran's Hospital, February 9, 1971, Box 2, Folder: San Fernando VA Hospital Earthquake, Donald Johnson Papers, Nixon Library. In same box and folder, see also "Subcommittee on San Fernando VA Hospital Disaster, Interim Report," March 1971.

50. *The Earthquake Disaster at the Veterans' Administration Hospital, San Fernando, Calif., on February 9, 1971: Hearings before a Special Subcommittee of the Committee on the Veterans' Affairs House of Representatives* (hereafter cited as *Veterans' Hospital Hearings*), 92 Cong. 169–71 (February 18, 1971) (statement of Los Angeles County county engineer). See also *Los Angeles Times*, February 10, 1971.

51. On structural failures of the hospital, see "Olive View Hospital Earthquake Report," Box 319, Folder: Earthquakes—1970–71, Hahn Collection, Huntington Library.

52. "Olive View Hospital Earthquake Report." See also *Veterans' Hospital Hearings* (statement of Henry Degenkolb, consulting structural engineer), 144–45.

53. Shirley Braverman and Nancy Jenks, "California Quake," *American Journal of Nursing* 71 (April 1971): 708–12.

54. Braverman and Jenks, "California Quake."

55. Braverman and Jenks, "California Quake."

56. Braverman and Jenks, "California Quake."

57. *California Earthquake Hearing* (statement of Reverend John G. Simmons, administrator, Pacoima Memorial Lutheran Hospital, 189–92). Also, James Thompson, "Damage to the Indian Hills Medical Building," in *San Fernando, California, Earthquake*, 198. See also *Los Angeles Times*, February 11 and 12, 1971.

58. *California Earthquake Hearing* (statement of Don Pool, manager, San Fernando Valley Human Resources Development Center, 239). Also, *Van Nuys News*, February 16, 1991.

59. The county set up special bus routes to help workers in Long Beach and Downey make the difficult commute. See *Van Nuys News*, February 26, 1971.

60. In 1971 the VA managed 166 hospitals and more than 250 smaller medical facilities. A few of the Sylmar employees took lower-paying jobs for the chance to work in specific facilities where there were no jobs at their grade level; a few also retired. See Memo titled "San Fernando Employees," n.d., Box 2, Folder: San Fernando VA Hospital Earthquake, Donald Johnson Papers, Nixon Library.

61. *New York Times*, February 17, 1971. Also, *Los Angeles Times*, February 11, 1971.

62. Press Release from the Office of Kenneth Hahn, February 9, 1971, Box 319, Folder: Earthquakes—1970–71, Hahn Collection, Huntington Library.

63. Brown had also served as secretary of the air force during the Vietnam War. After leaving Caltech in 1977, he served as the US secretary of defense. See Alice Stone's "Inter-

view with Harold Brown," March 19, 1990, in Caltech's Collection of Open Digital Archives, managed by Caltech Library Services, accessed June 16, 2020, http://oralhistories.library .caltech.edu/209/1/Brown,_Harold,_OHO.pdf.

64. "Presentation by Dr. Clarence R. Allen," March 10, 1971, Box 220, Folder: Earthquake Commission, 1971, Hahn Collection, Huntington Library.

65. *Van Nuys News*, March 4, 1971.

66. "Report on Olive View Hospital, May 25, 1971," by the Structural Engineers Association of Southern California, Box 319, Folder: Earthquakes—1970–71, Hahn Collection, Huntington Library. The board also received support from the recently established Earthquake Engineering Research Institution at UCLA, which was headed by C. Martin Duck and maintained a staff of one hundred persons.

67. J. Marx Ayres to W. T. Wheeler, April 21, 1971, Box 319, Folder: Earthquakes—1970–71, Hahn Collection, Huntington Library.

68. Report of Advisory Member of SEAOC, April 16, 1971, 4, Box 319, Folder: Earthquakes—1970–71, Hahn Collection, Huntington Library. In the same folder and file, see Don Wiltse to Harvey T. Brandt, June 4, 1971; and "Report on Olive View Hospital, May 25, 1971," by the Structural Engineers Association of Southern California.

69. *Report of the Los Angeles County Earthquake Commission, San Fernando Earthquake, February 9, 1971*, 1, Box 338, Folder: *Report of the Los Angeles Earthquake Commission*, Hahn Collection, Huntington Library, 1–2.

70. *Report of the Los Angeles County Earthquake Commission*, 3, 24–25.

71. Harvey Brandt, Country Engineer, to the US Senate Commerce Committee Subcommittee of Oceans and Atmosphere, n.d., Box 319, Folder: Earthquakes—1972–74, Hahn Collection, Huntington Library.

72. *Chief Administrative Officer's Progress Report to the Los Angeles County Board of Supervisors*, June 1973, Box 319, Folder: Earthquakes—1972–74, Hahn Collection, Huntington Library.

73. The committee had an equal number of members from the state senate and the assembly as well as the Democratic and Republican parties, and it represented communities across the state. For more on the committee's origins and achievements, see Geschwind, *California Earthquakes*, 173–75.

74. Geschwind, *California Earthquakes*, 173–75.

75. The original hospital bill placed the problem of seismic safety under the Office of Architecture and Construction. See Geschwind, *California Earthquakes*, 176–77.

76. On Alquist and Senate Bill 519, see press statement from the Office of Alfred E. Alquist, November 29, 1972, Box 7, Folder 362, in Alfred E. Alquist Papers, San Jose State University, San Jose, California. In same box and folder, see also Senator Alfred Alquist to Governor Ronald Reagan, November 10, 1972; and "SB 519 Statement," n.d.

77. Geschwind, *California Earthquakes*, 172, 177–78.

78. See "Suggested Interim Guidelines for the Seismic Safety Element in the General Plans," July 1972, Box GO 128, Folder: Intergovernmental Relations—Seismic Safety Element, Ronald Reagan Presidential Library and Museum, Simi Valley, California (hereafter cited as Reagan Library). See also James A. R. Johnson to City and County Administrators, July 17, 1972, Box GO 128, Folder: Intergovernmental Relations—Seismic Safety Element,

Reagan Library. Local administrators reacted to these initiatives in different ways. As a result, some communities were better protected from seismic hazards than others. See Geschwind, *California Earthquakes*, 179–80.

79. For more on California's seismic safety movement, including efforts to restrict construction near fault lines, see Geschwind, *California Earthquakes*, 180–84.

80. Lincoln to Richard Nixon, February 11, 1971, Subject Files (Disasters), Box 5, Folder: 1/1/71–2/28/71 [California], Nixon Library. The Office of Emergency Preparedness developed out of postwar agencies involved in civil defense planning, and it still had the responsibility of ensuring the safety of defense facilities.

81. Richard Nixon to Ronald Reagan, February 9, 1971, Subject Files (Disasters), Box 5, Folder: 1/1/71–2/28/71 [California], Nixon Library.

82. "Opening Remarks by Congressman Don Edwards, Chairman Special Subcommittee," n.d., Box 2, Folder: San Fernando VA Hospital Earthquake, Donald Johnson Papers, Nixon Library.

83. *Veterans' Hospital Hearings* (statement of Bruce Bolt, professor of seismology, 156–69).

84. *Veterans' Hospital Hearings* (statement of George Housner, earthquake engineer, 181–83).

85. *Veterans' Hospital Hearings* (statement of Henry Degenkolb, consulting structural engineer, 140–56).

86. "Subcommittee on the San Fernando VA Hospital Disaster: Interim Report," March 18, 1971, Box 2, Folder: San Fernando VA Hospital Earthquake, Donald Johnson Papers, Nixon Library.

87. On VA reforms, see B. A. Bolt, R. G. Johnston, J. Lefter, and M. A. Sozen, "The Study of Earthquake Questions Related to Veterans Administration Hospital Facilities," *Bulletin of the Seismological Society of America* 65, no. 4 (August 1975): 937–49.

88. See "Statement of Donald Johnson," January 12, 1972, Box 8, Folder: Hospitalization Rehabilitation, Donald Johnson Papers, Nixon Library. See also "Jeb Magruder to Mr. Haldeman," February 11, 1971, Box 5, Folder: DI 2/St 5 [California], White House Central Files, Disasters, Nixon Presidential Materials, Nixon Library.

89. "Statement of Donald Johnson."

90. Bolt et al., "Study of Earthquake Questions," 937–49.

91. For a summary of Cranston's role in this movement, see Thomas A. Birkland, *After Disaster: Agenda Setting, Public Policy, and Focusing Events* (Washington, DC: Georgetown University Press, 1997), 69–70; and Geschwind, *California Earthquakes*, 208–10.

92. The quake in China alone apparently claimed more than 240,000 lives. Birkland, *After Disaster*, 69–70; Geschwind, *California Earthquakes*, 208–10.

93. The 1994 quake occurred along a previously unidentified fault. The losses include indirect costs associated with lost business productivity and tax revenue as well as lost home equity and other costs that are difficult to calculate accurately. See *The Northridge Earthquake: Extent of Damage and Federal Response. Hearing before the Committee on Public Works and Transportation*, 103rd Cong., 2nd sess. (April 11, 1994), 6–7. Also, "Report to the Governor Pete Wilson," by Seismic Safety Commission, 1995, 1–5, Oviatt Library, California State University, Northridge.

94. The 1994 earthquake injured approximately nine thousand people, about sixteen hundred of them being admitted to hospitals. At least ten of the quake's victims died from stress-induced heart failures. See "Northridge Earthquake, January 17, 1994: Preliminary Reconnaissance Report," by the Earthquake Engineering Research Institute, 67–87, March 1994, Oviatt Library, California State University, Northridge.

95. The 1994 quake spawned its own new regulations. Twenty-three days after the quake, Governor Pete Wilson announced the establishment of another seismic safety commission. Working with more than thirty public and private organizations, the commission eventually produced a 135-page report with 168 new recommendations. The recommendations included new proposals to regulate the real estate and insurance industries, and to change the state's professional license renewal process. The commission also called for greater public funding for seismic safety research as well as new tax incentives to encourage homeowners to make their properties more earthquake resistant. California gradually adopted many of these proposals. See "The Northridge, California Earthquake, Risk Management Solutions 10-Year Retrospective," May 13, 2004, http://forms2.rms.com/rs/729 -DJX-565/images/eq_northridge_ca_eq.pdf.

96. In the words of California health officials, "Post-1973 hospital buildings and other health care facilities constructed under the requirements of the Seismic Safety Act performed very well with respect to the primary structural systems." See "Report to the Governor Pete Wilson," by Seismic Safety Commission, 1995, 67–73, Oviatt Library, California State University, Northridge.

97. On fear and modern hazards, see Ulrich Beck, *Risk Society: Towards a New Modernity* (London: Sage, 1992), 74–76. On the qualities of policy brokers, see John W. Kingdon, *Agendas, Alternatives, and Public Policies*, 2nd ed. (New York: Longman, 1994), 61–64, 189–90. See also Birkland, *After Disaster*, 18–19, 52–53, 62–72. On policy entrepreneurs and collective action, see Edella Schlager, "A Comparison of Frameworks, Theories, and Models of Policy Processes," in *Theories of the Policy Process*, ed. Paul Sabatier (Boulder, CO: Westview Press, 1999), 244–45. On the fleeting character of fights against nature, see Kelman, "Nature Bats Last," 391–402.

Chapter 5. MGM Grand Burns in Las Vegas!

Epigraph. Marshall Berman, *All That Is Solid Melts into Air: The Experience of Modernity* (New York: Penguin, 1982), 101.

1. *Los Angeles Times*, November 22, 1980. There were approximately one thousand people in the resort at the time of the fire.

2. *Los Angeles Times*, November 22, 1980.

3. A set designer who helped produce the resort's stage show had an assistant who also died in the fire, but the individual was not a hotel employee. See Deirdre Coakly, *The Day the MGM Grand Hotel Burned* (Secaucus, NJ: Lyle Stuart, 1982), 118.

4. Interview with Patricia Frymire, March 10, 2008. Frymire, a switchboard operator at the MGM, was on duty the day the resort caught fire.

5. Interview with Patricia Frymire, March 10, 2008.

6. On resort workers in Las Vegas, see James P. Kraft, *Vegas at Odds: Labor Conflict in a Leisure Economy, 1960–1985* (Baltimore: Johns Hopkins University Press, 2010). Also, Hal K.

Rothman and Mike Davis, eds., *The Grit beneath the Glitter: Tales from the Real Las Vegas* (Los Angeles: University of California Press, 2002); and Hal Rothman, *Neon Metropolis: How Las Vegas Started the Twenty-First Century* (New York: Routledge, 2002).

7. On the background and significance of the magnesium plant, see Gerald D. Nash, *World War II and the West: Reshaping the Economy* (Lincoln: University of Nebraska Press, 1990), 122–26. On urban developments in the West, see Carl Abbott, *The Metropolitan Frontier: Cities in the Modern American West* (Tucson: University of Arizona Press, 1993), 3–29.

8. See Robert D. McCracken, *Las Vegas: The Great American Playground* (Reno: University of Nevada Press, 1996), 64–65, 77–78; Kraft, *Vegas at Odds*, 9–32.

9. In 1966 and 1967, Nevada passed two acts that amended the state's gaming codes and opened gaming to corporate investments. See Kraft, *Vegas at Odds*, 26–28. For another overview of the period, see *Las Vegas Perspective: 1981*, 3–26, Special Collections, Lied Library, University of Nevada, Las Vegas (hereafter cited as UNLV Special Collections). The *Perspective* is a collection of facts and statistics compiled by the Nevada Development Authority, the First Interstate Bank of Nevada, and the Center for Business and Economic Research at the University of Nevada, Las Vegas. See also Hal Rothman, "Colony, Capital, and Casino: Money in the Real Las Vegas," in Rothman and Davis, *Grit beneath the Glitter*, 327.

10. Ramada Inns, *1979 Annual Report* (Phoenix: Ramada Inns, 1979), 12, Tropicana Promotional and Publicity Files, UNLV Special Collections; *Las Vegas Perspective: 1981*, 42–45, UNLV Special Collections. Other properties that opened off the Strip include Alexis Park, the Bingo Palace (now Palace Station), and Vegas World.

11. See Kraft, *Vegas at Odds*, 34, 140, 158–60; *Las Vegas Perspective: 1981*, 3–26, UNLV Special Collections. The airport was named after Nevada Senator Patrick McCarran, who played a prominent role in bringing new businesses to Nevada.

12. On working in resorts, see Kraft, *Vegas at Odds*, 33–52.

13. David G. Schwartz, *Suburban Xanadu: The Casino Resort on the Las Vegas Strip and Beyond* (New York: Routledge, 2003), 156, 163, 169–70. McCracken, *Las Vegas*, 95.

14. The only sprinklers in the hotel towers were on the top floor in an area designed for high-stakes gambling. The fire spread so rapidly that it disabled the hotel's manual fire alarm system. "Report of Clark County Fire Department, 1981," Section V, 14–18, accessed May 29, 2020, https://firerescuemagazine.firefighternation.com/wp-content/uploads/content/dam/fe/downloads/FFN-FRM-Downloads-Editorial/MGM_FIRE.pdf.

15. "Report of Clark County Fire Department," Section IV, 1–6.

16. *Las Vegas Sun* (hereafter cited as *Sun*), November 23, 1980.

17. *Sun*, November 23, 1980.

18. "Report of Clark County Fire Department," Section V, 7–18. Also, *Sun*, November 22, 1980; *San Francisco Chronicle*, November 23, 1980; and *Las Vegas Review-Journal* (hereafter cited as *Review-Journal*), November 19, 2000.

19. National Fire Protection Association, *Investigation Report on the MGM Grand Fire*, November 21, 1980 (revised January 15, 1982), 16, UNLV Special Collections (hereafter cited as *NFPA Investigation*).

20. "Report of Clark County Fire Department," Section IX, Statement from Tim Connor, 2–6.

21. "Report of Clark County Fire Department," Statement from Harvey B. Ginsberg, 17.

22. *Review-Journal*, November 22, 1980.

23. *Review-Journal*, November 22, 1980.

24. "Report of Clark County Fire Department," Section IX, Statement of Bobby Combs, 11.

25. "Report of Clark County Fire Department," Statement from Louis Miranti, 20–21.

26. *Los Angeles Times*, November 23, 1980.

27. "A Story about the MGM Fire," Iklim, March 5, 2008, http://iklimnet.com/hotelfires /caseMGMstory1_2.html.

28. On fire service response, see *NFPA Investigation*, 21–27; and "Report of Clark County Fire Department," Section IX, Statement of Bert Sweeny, 22–23.

29. *NFPA Investigation*, 26–27; "Report of Clark County Fire Department," Section IX, Statement of Bert Sweeny, 22–23.

30. *Sun*, November 22, 1980.

31. *Sun*, November 22, 1980.

32. *Sun*, November 22, 1980.

33. Interview with Charles Blake, December 29, 2009, Las Vegas. Blake worked at the Las Vegas office of the Federal Bureau of Investigation and volunteered his services to the Red Cross during and after the MGM fire.

34. About half of the ironworkers who participated in rescue efforts were treated for smoke inhalation, and several had to be hospitalized. See Coakly, *The Day the MGM Grand Hotel Burned*, 75–77; *Sun*, November 23, 1980.

35. *Los Angeles Times*, November 23, 1980.

36. Coakly, *The Day the MGM Grand Hotel Burned*, 117.

37. Interview with Sherry McClaren Walberg, June 17, 2009, Las Vegas. Walberg was working as a cashier at the MGM when the fire broke out. See also *Los Angeles Times*, November 22 and 23, 1980; *Sun*, November 23, 1980. Guests reported the loss of more than $270 million in cash, jewelry, and other possessions as a result of the fire (worth $840 million in 2020). According to the MGM, the only money it lost in the casino was a relatively small amount of cash in the keno lounge. On theft during the fire, see *Sun*, November 24, 1980.

38. *San Francisco Chronicle*, November 23, 1980.

39. *San Francisco Chronicle*, November 24, 1980.

40. *San Francisco Chronicle*, November 24, 1980. On the role of the media after disasters, see John W. Kingdon, *Agendas, Alternatives, and Public Policies*, 2nd ed. (New York: Longman, 1994), 61–64; Thomas A. Birkland, *Lessons of Disaster: Policy Change after Catastrophic Events* (Washington, DC: Georgetown University Press, 2007), 26–27, 173–74.

41. *Review-Journal*, December 3, 1980, and April 28, 1983. The state unemployment rate stood at 6.7 percent in October 1980, compared to the national rate of 7.6 percent. The closure of the Aladdin in the summer of 1980 partly explains the high unemployment in Clark County. The Nevada State Gaming Commission shut down the Aladdin for three months in 1980, from July to October, after learning that unlicensed people were running its casino in violation of state gaming laws. The shutdown left two thousand people out of work. On the history of the Aladdin and the gaming commission, see *Review-Journal*, July 11, 1980.

42. On employment in resorts in these years, see *Las Vegas Perspective: 1981*, 39; and *Las Vegas Perspective: 1985*, 76, UNLV Special Collections. Also, "Nevada Economic Indicators," Papers of Governor Robert List, Folder: Labor, Box 1180, Folder 046, Nevada State Museum and Archives, Special Collections, Carson City, Nevada (hereafter cited as NSMA Special Collections).

43. Interview with Kathy Griggs, April 24, 2006, Las Vegas.

44. *San Francisco Chronicle*, November 24, 1980. Interview with Patricia Frymire, March 10, 2008, Las Vegas.

45. *Review-Journal*, November 25, 1980.

46. Interview with Kathy Griggs, April 24, 2006, Las Vegas.

47. On Local 226 at the time of the fire, see Senate Committee on Governmental Affairs, *Hotel Employees and Restaurant Employees International Unions*, 98th Cong., 2nd sess. (1984), 68.

48. *Review-Journal*, November 28, 1980. To stay in shape after the fire, some of the displaced dancers enrolled in exercise classes.

49. *Review-Journal*, November 28, 1980. For more on Ford's experience, see *Review-Journal*, July 31, 1981.

50. Interview with Jim Bonaventure, May 5, 2006, Las Vegas. The maximum benefits the state paid to MGM workers amounted to $127 a week (about $400 in 2020).

51. Interview with Kathy Griggs, April 24, 2006, Las Vegas.

52. Interview with Patty Jo Allsbrook, March 14, 2008, Las Vegas.

53. Committee on Human Resources and Facilities, Sixty-First Session, Nevada State Legislature, "Senate Bill No. 214, February 11, 1981" (statement of Senator William H. Hernstadt, 50), accessed June 17, 2020, http://www.leg.state.nv.us/Division/Research/Library/LegHistory/LHs/1981/SB214,1981pt1.pdf.

54. *New York Times*, March 13, 1981; *Sun*, January 25, 2008. On the governor's commitment to minimizing the fire's impact, see his letter to Harriet Copeland, December 29, 1980, Papers of Governor Robert List, Box 0837, Folder: Fire, NSMA Special Collections.

55. The arsonist, twenty-three-year-old busboy Philip Cline, was convicted in 1982 of arson and eight counts of murder, for which he received a sentence of life imprisonment. For more on Cline, see *Review-Journal*, February 6, 2011. On the fire, see *Sun*, February 11, 1981. See also *New York Times*, March 26, 1982, and October 20, 1985.

56. The MGM Grand eventually paid out more than $167 million to plaintiffs as a result of the fire, and Hilton paid $16.5 million. *Las Vegas Sun*, February 11, 1981; *Los Angeles Times*, April 1, 1985. The latter article discussed the MGM lawsuit against its fire insurance companies, which was a product of the fire.

57. On risk bearers and risk generators, see Ingar Palmlund, "Social Drama and Risk Evaluation," in *Social Theories of Risk*, ed. Sheldon Krimsky and Dominic Golding (Westport, CT: Praeger, 1992), 202–6.

58. See editorial by state senator Joe Neal, who introduced the senate version of the fire safety legislation, in *Review-Journal*, January 23, 2006. On the initial version of Neal's bill, see Committee on Human Resources and Facilities, "Senate Bill No. 214," 3–8.

59. On the proposed powers and responsibilities of the Board of Fire Safety, see Committee on Human Resources and Facilities, "Senate Bill No. 214," 6–8.

60. Nevada State Assembly Committee on Government Affairs, "Minutes of the Nevada State Legislature," May 6, 1981 (statement of Vern Balderston, 144–46), https://www.leg.state.nv.us/Division/Research/Library/LegHistory/LHs/1981/SB214,1981pt2.pdf.

61. John L. Smith, *The Westside Slugger: Joe Neal's Lifelong Fight for Social Justice* (Reno: University of Nevada Press, 2019), 133–39. See also the oral history by Dana R. Bennett, *John E. "Jack" Jeffrey* (Carson City, NV: Legislative Counsel Bureau, 2009), https://www.leg.state.nv.us/Division/Research/LegInfo/OHP/transcripts/Jeffrey.pdf.

62. On the expansion of resorts in these years, see Kraft, *Vegas at Odds*, 9–32.

63. "MGM Spends $5 Million on Fire Safety," UPI Archives, March 3, 1981, https://www.upi.com/Archives/1981/03/03/MGM-spends-5-million-on-fire-safety/1143352443600/.

64. "Minutes of Nevada State Legislature," May 6, 1981 (statement of E. A. Schweitzer, 154–58).

65. "Minutes of Nevada State Legislature," May 6, 1981 (statement of Robbins Cahill, 154–56).

66. *Los Angeles Times*, July 30, 1980.

67. Nearly two thousand sprinkler heads were installed in the ceiling of the MGM's newly expanded casino. *Reno Evening Gazette*, July 30, 1981; *Nevada Appeal*, July 30, 1981.

68. *Valley Times*, July 30, 1981; *Los Angeles Times*, July 30, 1981.

69. Interview with Cathy Beane, June 17, 2009, Las Vegas.

70. These are the words of Rosalie Manganelli, a hostess in the MGM delicatessen who narrowly escaped the fire, as quoted in Coakly, *The Day the MGM Grand Hotel Burned*, 189–91. Jim Bonaventure, Kathy Griggs, and Patty Jo Allsbrook returned to work at the MGM.

71. *Review-Journal*, July 30 and 31, 1981.

72. Interview with Eileen Daleness, June 17, 2009, Las Vegas.

73. Florida's ordinances applied only to lodging establishments with interior hallways. The US Fire Administration compiled and published the list of buildings that had sprinkler systems. See Gary Stoller, "Better Fire Safety in Hotels Saves Lives," *USA Today*, November 20, 2005, http://www.ips-blazemaster.com/images/pdfs/USAToday2005-11-20.pdf.

74. Stoller, "Better Fire Safety in Hotels Saves Lives." See also Richard Campbell, *Hotel and Motel Structure Fires* (Quincy, MA: National Fire Protection Association, September 2015), https://www.nfpa.org/News-and-Research/Data-research-and-tools/Building-and-Life-Safety/Hotel-and-Motel-Structure-Fires; and Institute for Real Estate Management, "Life Safety Laws: Sprinklers in High-Rise Buildings," accessed November 10, 2015, https://silo.tips/download/sprinklers-in-high-rise-buildings.

75. The problem of fire safety looms much larger in other parts of the world, including Europe. Only a small percentage of European hotels have automatic sprinkler systems, and many of those places do not have sprinklers in every room. Sprinkler requirements in the European Union still vary considerably from nation to nation, and even from city to city. *USA Today*, November 20, 2005. Two reports produced for the European Union offer insights into the subject: Blanca Ballester, *Hotel Fire Safety: The Case for Legislation* (Brussels: Secretariat of the European Parliament, June 2013), https://www.europarl.europa.eu/RegData/etudes/note/join/2013/504470/IPOL-JOIN_NT(2013)504470_EN.pdf; and European Association for the Co-ordination of Consumer Representation in Standardization (ANEC), *Hotel Fire Safety in the European Union: The Consumer Perspective* (Brussels: ANEC, October

2010), https://www.yumpu.com/en/document/view/25104106/hotel-fire-safety-in-the-euro
pean-union-anec.

Chapter 6. Federal Building Bombed in Oklahoma City!

Epigraph. Friedrich Nietzsche, *Beyond Good and Evil: Prelude to a Philosophy of the Future,*
trans. Marion Faber (Oxford: Oxford University Press, 1998), 3.

1. On McVeigh's justification for the bombing, see "McVeigh's Letter to Fox News,"
April 23, 2001, www.outpost-of-freedom.com/mcveigh/okcwhy.htm. "When an aggressor
force continually launches attacks from a particular base of operation, it is sound military
strategy to take the fight to the enemy," McVeigh wrote. As authorities soon learned, Michael Fortier and his wife, Lori, knew about the nefarious plan but failed to inform them.
On the number and types of injuries, see the report funded by the Oklahoma State Department of Health: Sheryll Shariat, Sue Mallonee, and Shelli Stephens Stidham, *Oklahoma City
Bombing Injuries* (Oklahoma City: Oklahoma State Department of Health, December 1998),
1–6, https://www.ok.gov/health2/documents/OKC_Bombing.pdf. One responding emergency worker died as a result of the bombing and was listed among the 168 fatalities.

2. For an inside look at the hunt for the perpetrators, see John Hersley, Larry Tongate,
and Bob Burke, *Simple Truths: The Real Story of the Oklahoma City Bombing Investigation*
(Oklahoma City: Heritage Association, 2004).

3. Useful books on the bombing include Lou Michel and Dan Herbeck, *American Terrorist: Timothy McVeigh and the Oklahoma City Bombing* (New York: Harper, 2001); Stuart A.
Wright, *Patriots, Politics, and the Oklahoma City Bombing* (New York: Cambridge University
Press, 2007); Edward T. Linenthal, *The Unfinished Bombing: Oklahoma City in American Memory* (New York: Oxford University Press, 2001).

4. On the bombing and the media, see James David Ballard, *Terrorism, Media and Public
Policy: The Oklahoma City Bombing* (New York City: Hampton Press, 2005). On political as-
pects of the tragedy, see David Hoffman, *The Oklahoma City Bombing and the Politics of Terror* (Venice, CA: Feral House, 1998). For other scholarly perspectives of the bombing, see
Chris Piotrowski and Bob Perdue, "Researching the Oklahoma City Bombing," *Psychological
Reports* 77, no. 3 (December 1, 1995): 1099–1102. On emergency workers and the bombing,
see P. A. Maningas, M. Robison, and S. Mallonee, "The EMS Response to the Oklahoma City
Bombing," *Prehospital Disaster Medicine* 24, no. 2 (June 1997): 9–14. On the role of government agencies, see Alethia H. Cook, "The 1995 Oklahoma City Bombing: Bureaucratic Response to Terrorism and a Method for Evaluation" (PhD diss., Kent State University, 2006).
On the lugubriously named Antiterrorism and Effective Death Penalty Act of 1996 (AEDPA),
see John H. Blume, "AEDPA: The 'Hype' and the 'Bite,'" *Cornell Law Review* 91, no. 2 (January 2006): 259–302.

5. These figures are drawn from the *Oklahoma City National Memorial Factbook*, a work
compiled by the memorial's staff (and hereafter cited as *OKC Memorial Factbook*) that is
located in the Oklahoma City National Memorial Archives (hereafter cited as OKC Memorial Archives). The book says 118 working people died in the Murrah Building, but one is
described as "company representative making a delivery."

6. For an overview of the city's history, see *Encyclopedia of Oklahoma History and Culture*, accessed May 31, 2020, https://www.okhistory.org/publications/encyclopediaonline.

7. *Encyclopedia of Oklahoma History and Culture.*

8. Richard M. Bernard, "Oklahoma City: Booming Sooner," in *Sunbelt Cities: Politics and Growth since World War II*, ed. Richard M. Bernard and Bradley R. Rice (Austin: University of Texas Press, 1983), 213–34.

9. Lynda S. Ozan, *The Historic Context for Modern Architecture in Oklahoma: Housing from 1946–1976* (Washington, DC: National Park Service, October 2014), http://www.ok history.org/shpo/thematic/modernarchitecture.pdf.

10. "Final Report: Alfred P. Murrah Federal Building Bombing," 2–5, OKC Memorial Archives; *The Daily Oklahoman*, June 19, 1995 (the paper was renamed *The Oklahoman* in 2003, and it is hereafter cited by that name).

11. *The Oklahoman*, October 22, 1974. See also "Arrangements for Groundbreaking Ceremony," October 21, 1971, in "Alfred P. Murrah Building Dedication Ceremony," Collection 474, Box 001, OKC Memorial Archives. President Franklin Roosevelt nominated Murrah as a US District Judge in 1937. When Congress confirmed the nomination later that year, Murrah became one of the youngest federal judges in American history.

12. *The Oklahoman*, October 22, 1974.

13. *The Oklahoman*, October 22, 1974. See also *OKC Memorial Factbook*. A computerized heating and cooling system in the building regulated temperatures to conserve energy.

14. See "Official Dedication Ceremony, News Release," Collection 474, Box 001, OKC Memorial Archives.

15. Interview with Richard Williams, March 28, 2012. The GSA had been an independent agency within the government since 1949, when a Congress intent on streamlining administrative activities combined the responsibilities of several federal agencies.

16. Interview with Richard Williams, March 28, 2012.

17. See "Alfred P. Murrah Federal Building," Collection 474, Box 001, OKC Memorial Archives.

18. Interview with Richard Williams, November 1, 2011. As Williams explained, the FPS control center was located in the basement of the downtown federal courthouse. See also Shawn Reese and Lorraine H. Tong, *Federal Building and Facility Security* (Washington, DC: Congressional Research Service, March 24, 2010), 6, https://apps.dtic.mil/dtic/tr/fulltext /u2/a517334.pdf.

19. *New York Times*, October 23 and December 20, 1983. See also "Beirut Marine Barracks Bombing Fast Facts," CNN, November 13, 2019, https://www.cnn.com/2013/06/13 /world/meast/beirut-marine-barracks-bombing-fast-facts/index.html.

20. US Navy Admiral Bobby Ray Inman chaired the panel. Formally known as the Secretary of State's Advisory Panel on Overseas Security, the panel released its report in 1985. See *The Inman Report: Report of the Secretary of State's Advisory Panel on Overseas Security* (Washington, DC: US Department of State, 1985), https://fas.org/irp/threat/inman/.

21. 104 Cong. Rec. 11537 (May 2, 1995).

22. On the size of specific agencies in the building, see information about injuries and diagrams of the structure in *OKC Memorial Factbook*.

23. *OKC Memorial Factbook.*

24. Interview with Dianne Dooley, December 8, 2011.

25. *Wall Street Journal*, April 26, 1995; Interview with Florence Rogers, October 13, 2011.

The FECU had moved into the building from the federal courthouse in 1977, when it was a considerably smaller enterprise.

26. Melva Noakes owned the day-care center. For more on the center and the recollections of its owner, see *New York Times*, May 3, 1995.

27. Shariat et al., "Oklahoma City Bombing Injuries," December 1998, 1–5.

28. On the relationship between location and injuries, see "Investigation of Physical Injuries Directly Associated with the Oklahoma City Bombing," Collection 246, Box B3, Folder F2, OKC Memorial Archives.

29. Lorri McNiven, Survivors' Recollections, Survivors' Recollections Binder (hereafter cited as Survivors' Recollections), OKC Memorial Archives.

30. Army Recruiter Henderson Baker, Survivors' Recollections, OKC Memorial Archives.

31. Donald W. Bewley, Survivors' Recollections, OKC Memorial Archives.

32. Priscilla Salyers, Survivors' Recollections, OKC Memorial Archives. Salyers's story is also retold in *Ladies' Home Journal* (April 1996): 200–202, Collection 246, Box B2, Folder F118, OKC Memorial Archives.

33. Interview with Florence Rogers, April 13, 2011.

34. Interview with Florence Rogers, April 13, 2011; *Wall Street Journal*, April 26, 1995.

35. Shariat et al., "Oklahoma City Bombing Injuries," 2–5.

36. Marsha Kight, *Forever Changed: Remembering Oklahoma City, April 19, 1995* (Amherst, NY: Prometheus Books, 1998), 70–72.

37. Kight, *Forever Changed*, 146–50.

38. Interview with Dianne Dooley, October 17, 2011.

39. Interview with Dianne Dooley, October 17, 2011.

40. Kight, *Forever Changed*, 150.

41. Kight, *Forever Changed*, 54.

42. Dot Hill, Survivors' Recollections, OKC Memorial Archives.

43. Interview with Janet Beck, October 25, 2011.

44. *The Oklahoman*, April 20, 1995. See also Rev. Boyce A. Boyden, "Oklahoma City Bombing Survivor Says God Helped Her Forgive," United Methodist News Service, April 14, 2015, https://www.umnews.org/en/news/oklahoma-city-bombing-survivor-says-god-helped -her-forgive.

45. Boyden, "Oklahoma City Bombing Survivor."

46. Kight, *Forever Changed*, 135–38.

47. Kight, *Forever Changed*, 226.

48. Kight, *Forever Changed*, 173–77; Interview with Dot Hill, October 18, 2011. For more on religious beliefs and the bombing, see Kenneth I. Pargament, Bruce W. Smith, Harold G. Koenig, and Lisa Perez, "Patterns of Positive and Negative Religious Coping with Major Life Stressors," *Journal for the Scientific Study of Religion* 37, no. 4 (1998): 711–25.

49. *New York Times*, April 24, 1995; "Reverend Billy Graham: Oklahoma Bombing Memorial Prayer Service Address," American Rhetoric Online Speech Bank, April 23, 1995, https://www.americanrhetoric.com/speeches/billygrahamoklahomabombingspeech.htm.

50. "How Can You Mend a Broken Heart" was written and composed by Barry and Robin Gibb of the Bee Gees. They released it in 1971.

51. Interview with Jennifer Takagi, October 17, 2011.

52. Interview with Beverly Rankin, October 24, 2011.

53. Charles E. Fritz, "Disaster," in *Contemporary Social Problems*, ed. Robert K. Merton and Robert A. Nisbet (New York: Harcourt, Brace and World, 1961), 683–84. On the solidarity of disaster victims, see Jacob A. C. Remes, *Disaster Citizenship: Survivors, Solidarity, and Power in the Progressive Era* (Urbana: University of Illinois Press, 2016), 45–48, 70–71.

54. About three hundred of the Medallion Hotel's four hundred rooms were canceled within hours of the bombing, and subsequently most of them were then reserved for newspaper and television crews. See *New York Times*, April 24, 1995.

55. Email correspondence with Richard Williams, July 12, 2012.

56. Email correspondence from Germaine Johnston, October 20, 2011.

57. Email correspondence from Germaine Johnston, October 20, 2011.

58. Email correspondence from Germaine Johnston, October 20, 2011.

59. "Recovery of SSA Employees after Murrah Building Bombing," December 4, 1995. Factsheet on post-bombing developments within the Social Security Administration, provided to author by Dennis Purifoy, October 22, 1995.

60. Interview with Beverly Rankin, October 24, 2011.

61. Email correspondence with Janet Beck, October 20, 2011.

62. Interview with Beverly Rankin, October 24, 2011.

63. Grayson R. Towler, "A Survivor's Tale: The Rebirth of the Federal Employees Credit Union in Oklahoma City," *Disaster Recovery Journal* 8, no. 3 (July/August/September 1995): 11–14.

64. Interview with Florence Rogers, October 13, 2011; email correspondence with Rogers, October 5, 2011.

65. Interview with Jennifer Takagi, October 17, 2011. The proximity of employees in the old Montgomery Ward store may have complicated Takagi's own recovery. "We were all crammed together," she recalled.

66. Stephanie Smith, *The Interagency Security Committee and Security Standards for Federal Buildings* (Washington, DC: Congressional Research Service, November 23, 2007), 1–5, www.fas.org/sgp/crs/homesec/RS22121.pdf. The president established the ISC in October 1995.

67. 104 Cong. Rec. 11540 (May 2, 1995).

68. *Extensions of Remarks*, 104 Cong. Rec. 15992 (June 27, 1996). For a deeper understanding of such debates, see remarks of senators about the proposed Comprehensive Terrorism Prevention Act, 104 Cong. Rec. 7656–7690 (June 5, 1995); 104 Cong. Rec. 14932–14961 (June 6, 1996); 104 Cong. Rec. 7718–7758 (June 6, 1995); and 104 Cong. Rec. 9361–9364 (April 29, 1996).

69. See *Oklahoma City Federal Building: A National Symbol of Strength and Renewal*, 6–11, OKC Memorial Archives. On the National Memorial and Museum, which opened on the seventh anniversary of the bombing, see the museum's brochure *1995–2005, a Decade of Hope: Oklahoma City National Memorial* (Oklahoma City: Oklahoma City National Memorial Foundation, 2005), 43–48.

70. *The Oklahoman*, May 4, 2004. Also, *Oklahoma City Federal Building*, 8–21; Eve E. Hinman, "Security Management: Designing Blast Resistant Buildings: Lessons Learned from Oklahoma City," Collection 246, Box B6, Folder F158, OKC Memorial Archives.

71. *The Oklahoman*, May 4, 2004.

72. Interview with Beverly Rankin, October 24, 2011.

73. Interview with Dot Hill, October 18, 2011.

74. Interview with Dot Hill, October 18, 2011. Also, interview with Jennifer Takagi, October 17, 2011.

75. Email with Richard Williams, August 22, 2012. "Survivors of some agencies emotionally could not and did not want to move back into a Federal Complex," Williams wrote. See also email with Germaine Johnston, August 22, 2012.

76. Interview with Jennifer Takagi, October 17, 2011.

77. "I liked it all," Dooley explained. Interview with Dianne Dooley, October 17, 2011.

78. Interview with Dianne Dooley, October 17, 2011.

79. Interview with Jennifer Takagi, October 17, 2011.

80. Report of US Customs Service, "The OKC Tragedy," Vol. 30, no. 2, in Collection 246, Box B3, Folder F2, OKC Memorial Archives.

81. *The Oklahoman*, May 11, 1995.

82. *The Oklahoman*, May 29, 1995.

83. *The Oklahoman*, May 29, 1995.

84. On historians' neglect of public workers, see Joseph A. McCartin, "Bringing the State's Workers In: Time to Rectify an Imbalanced U.S. Labor Historiography," *Labor History* 47, no. 1 (2006): 3–94. See also Robert Shaffer, "Where Are the Organized Public Employees? The Absence of Public Employee Unionism from U.S. History Textbooks, and Why It Matters," *Labor History* 43, no. 3 (2002): 315–35.

Conclusion

Epigraph. Kofi Annan, *We the Peoples: The Role of the United Nations in the 21st Century* (New York: United Nations Department of Public Information, 2000), 5.

1. Charles E. Fritz, "Disaster," in *Contemporary Social Problems*, ed. Robert K. Merton and Robert A. Nisbet (New York: Harcourt, Brace and World, 1961), 651–94; Kevin Rozario, *The Culture of Calamity: Disaster and the Making of Modern America* (Chicago: University of Chicago Press, 2007), 1–9, 22–23.

2. Gerald D. Nash, *The Federal Landscape: An Economic History of the Twentieth-Century West* (Tucson: University of Arizona Press, 1999), 41–99; Carl Abbott, *The Metropolitan Frontier: Cities in the Modern American West* (Tucson: University of Arizona Press, 1993), 3–29.

3. Roger W. Lotchin, "California Cities and the Hurricane of Change: World War II in the San Francisco, Los Angeles, and San Diego Metropolitan Areas," *Pacific Historical Review* 63, no. 3 (August 1994): 393–420; idem, *The Bad City in the Good War: San Francisco, Los Angeles, Oakland, and San Diego* (Bloomington: University of Indiana Press, 2003).

4. Gerald D. Nash, *The American West in the Twentieth Century: A Short History of an Urban Oasis* (Albuquerque: University of New Mexico Press, 1973), 193–203; idem, *World War II and the West: Reshaping the Economy* (Lincoln: University of Nebraska Press, 1990), 4, 42, 68–69, 131. For a more thorough investigation of Southern California during the war, see Roger W. Lotchin, *Fortress California, 1911–1960: From Warfare to Welfare* (New York: Oxford University Press, 1992).

5. On Sylmar's population growth, see articles about the community in the *Los Angeles Times*, May 6, 1962, and June 21, 1972.

6. On early postwar growth in the West, see Abbott, *Metropolitan Frontier*; Richard M. Bernard and Bradley R. Rice, *Sunbelt Cities: Politics and Growth since World War II* (Austin: University of Texas Press, 1983). Also, see Peter Wiley and Robert Gottlieb, *Empires in the Sun: The Rise of the New American West* (New York: G. P. Putnam's Sons, 1982).

7. Henry George, *Progress and Poverty: An Inquiry into the Cause of Industrial Depressions and of Increase of Want with Increase of Wealth* (New York: D. Appleton, 1879), 3–13; idem, *Social Problems* (New York: Henry George, 1886), 192–203.

8. Ulrich Beck, *Risk Society: Towards a New Modernity* (London: Sage, 1992), 19–27; George, *Progress and Poverty*, 8.

9. According to social theorist Anthony Giddens, "The development of modern social institutions and their worldwide spread have created vastly greater opportunities for human beings to enjoy a secure and rewarding existence than any type of pre-modern system." "But modernity also has a sombre side, which has become very apparent in the present century." *The Consequences of Modernity* (Stanford, CA: Stanford University Press, 1990), 7–10.

10. Eugene A. Rosa, Ortwin Renn, and Aaron M. McCright, *The Risk Society Revisited: Social Theory and Governance* (Philadelphia: Temple University Press, 2014), xxv–xxviii. As these authors say in the preface to their book: "But all scientific developments that drive technological change are accompanied by an unshakeable traveling companion: risk."

11. See Beck, *Risk Society*, 13; and Rosa et al., *Risk Society Revisited*, xvii.

12. On the significance of work in modern times, see Ulrich Beck, *Brave New World of Work* (Malden, MA: Blackwell, 2000), 7, 14, 63.

13. On disasters as focusing events, see Thomas A. Birkland and Megan K. Warnement, "Focusing Events, Risk, and Regulation," in *Policy Shock: Recalibrating Risk and Regulation after Oil Spills, Nuclear Accidents and Financial Crises*, ed. Edward J. Balleisen et al. (New York: Cambridge University Press 2017), 107–28. On regulations as *focusing devices*, see Lee Vinsel, *Moving Violations: Automobiles, Experts, and Regulations in the United States* (Baltimore: Johns Hopkins University Press, 2019), 301–4, 314.

14. On the role of elected officials in post-disaster policy making, see John W. Kingdon, *Agendas, Alternatives, and Public Policies*, 2nd ed. (New York: Longman, 1994), 23–47.

15. See John L. Smith, *The Westside Slugger: Joe Neal's Lifelong Fight for Social Justice* (Reno: University of Nevada Press, 2019), 139.

16. On technical experts and policy entrepreneurs, see Kingdon, *Agendas, Alternatives, and Public Policies*, 4, 21, 99–105, 176–79, 188–93. See also Paul A. Sabatier, ed., *Theories of the Policy Process* (Boulder, CO: Westview Press, 1999). On the role of technical experts in standard setting, see JoAnne Yates and Craig N. Murphy, *Engineering Rules: Global Standard Setting since 1880* (Baltimore: Johns Hopkins University Press, 2019), 7–14.

17. On media coverage of disasters, see Rozario, *Culture of Calamity*, 192–96.

18. The Baden Aniline and Soda Factory (BASF) was the world's largest chemical producer. See *New York Times*, October 31, 1921.

19. On the Beirut attacks, see *New York Times*, October 23, 1983. According to Ronald K. Noble, undersecretary of the US Treasury, in 1994 there were forty-four bomb threats made against the Internal Revenue Service; see *New York Times*, May 2, 1995.

20. On the "policy primeval soup," see Kingdon, *Agendas, Alternatives, and Public Policies*, 21, 101–4.

21. On the air accident rate since 1960, see Boeing, *Statistical Summary of Commercial Jet Airplane Accidents: Worldwide Operations, 1959–2016* (Seattle: Boeing, July 2017), http://www.boeing.com/resources/boeingdotcom/company/about_bca/pdf/statsum.pdf.

22. On the 1960 air accident, see Marion F. Sturkey, *Mid-Air: Accident Reports and Voice Transcripts from Military and Airline Mid-Air Collisions* (Plum Branch, SC: Heritage Press International, 2008), 129–37. As a result of the collision, the FAA reduced the speed at which planes approached airports and required pilots to inform controllers whenever navigation and communications devices malfunctioned.

23. As quoted in an editorial that appeared in *Newsletter for Latin America and the Caribbean* 15 (1999): http://www.eird.org/eng/revista/No15_99/pagina1.htm.

24. On the importance of maintenance in modern society, see Andrew Russell and Lee Vinsel, "Let's Get Excited about Maintenance!," *New York Times*, July 23, 2017. See also idem, "Hail the Maintainers," *Aeon*, April 7, 2016.

25. Marshall Berman, *All That Is Solid Melts into Air: The Experience of Modernity* (New York: Penguin, 1988), 14–15, 23. The book, whose title draws from the words of Marx and Engels, is dedicated to the child Berman lost in a domestic tragedy.

26. See Berman, *All That Is Solid Melts Into Air*, 15–17, 21–23, 90, and 346–48. Eric Hobsbawm used the phrase "dialectical dance" to describe the shifting positions of moderate, middle-class reformers during the French Revolution. *The Age of Revolution, 1789–1848* (New York: Vintage Books, 1996), 62. (The work was originally published in 1962.)

Essay on Primary Sources

Materials relevant to the study of disasters and safety reforms stretch across literary and disciplinary boundaries. This essay directs readers to primary source materials that proved especially useful to this study. For information about the secondary sources consulted in the preparation of this book, readers should consult the book's plentiful end notes.

Archival collections form the bedrock upon which this book is built, and each of the book's five case studies stands upon a specific set of these sources. The study of the 1947 Texas City disaster relies largely on materials found at four research institutions in Texas. The National Archives in Fort Worth contains records of the US Court of Appeals for the Fifth Circuit, which has jurisdiction over appeals brought from the state of Texas and thus maintains records relevant to the court battles that arose over the disaster. Also in the Fort Worth archives are summaries of post-disaster interviews conducted by the Federal Bureau of Investigation, which help to underscore the disaster's impact on the waterfront community. The Woodson Research Center at Rice University in Houston, located within the Fondren Library, holds similarly valuable materials. Most importantly, it has records of the Monsanto Chemical Plant, including records regarding personnel matters. The Moore Memorial Public Library in Texas City also holds significant materials, such as official reports and unpublished studies about the Gulf Coast area and its economy. The library also has back copies of the *Texas City Sun*, whose reporters covered the waterfront tragedy in great detail. The University of Texas's Moody Medical Library in Galveston houses the Truman G. Blocker Jr. History of Medicine Collections, where researchers can find insights into the medical community's response to the disaster.

Researchers interested in the 1956 air disaster over the Grand Canyon should visit the Research Center of the State Historical Society of Missouri, located on the campus of the University of Missouri–Kansas City. Trans World Airlines maintained its corporate headquarters in Kansas City until 1964, and the company left many of its records with the Historical Society. Within those records are accident reports from the Civilian Aviation Administration as well as an assortment of press releases and newspaper clippings relevant to the disaster. The society also holds the Active Retired Pilots Association Collections, within which researchers can find useful copies of the association's trade journal, *The Airline Pilot*. The journal has much to say about the early history of the airline industry and the safety concerns of pilots. Of similar importance are the papers of Gordon R. Parkinson, a former TWA employee who served as president of the company's Seniors Club. Among other things, those papers contain correspondence relevant to emergency response and recovery efforts.

The study of the San Fernando Valley earthquake of 1971 rests largely on primary sources available at four research institutions in Southern California. The area's two presi-

dential libraries have a wealth of relevant documents about the quake. The Richard M. Nixon Presidential Library and Museum in Yorba Linda holds the papers of Donald Johnson, who headed the Veterans Administration during the Nixon years. Johnson's papers contain important information about the collapse of the Veterans Hospital in the San Fernando Valley and the agency's response to the crisis it created. The Ronald Reagan Library and Museum in Simi Valley also has a wide range of quake-related materials. As governor of California from 1967 to 1975, Reagan necessarily became involved in seismic safety campaigns, and his library's holdings help place those campaigns within their political context. Researchers can find a smorgasbord of additional resources in the Munger Research Center at the Huntington Library in Pasadena. Among the center's holdings are the papers of Kenneth Hahn, who served on the Los Angeles County Board of Supervisors from 1952 to 1992 and played a vital role in strengthening California's seismic safety codes. Hahn's correspondence with seismologists and engineers at the California Institute of Technology are particularly insightful. Researchers can find more quake-related resources at the California State Northridge Library. The library has a lengthy run of the *Van Nuys News*, whose stories often centered on how the 1971 quake affected the local community. The library also has official reports on the 1994 earthquake that rattled the San Fernando Valley, which help to evaluate the successes and failures of seismic safety regulations.

Primary sources relevant to the 1980 MGM Grand Hotel fire can be found at the Special Collections Department within the University of Nevada's Lied Library in Las Vegas. The department has a wide range of materials relevant to the rise of Las Vegas resorts, including the original MGM Grand Hotel. The Lied Library has an excellent collection of regional newspapers that covered the fire, including the *Las Vegas Review-Journal*, *Las Vegas Sun*, and *San Francisco Chronicle*. The Nevada State Museum and Archives in Carson City houses the papers of Robert List, who served as Nevada's governor from 1979 to 1983, and within those papers are documents about the causes and impact of the resort fire. The Research Division within the Nevada Legislative Council Bureau in Carson City can help researchers find records germane to the state's fire safety movement. The Nevada State Historical Society in Reno has additional materials of interest, although the materials are more about the early history of Nevada than the modern history of Las Vegas.

Those interested in the 1995 Oklahoma City bombing should visit the Oklahoma City National Memorial and Museum, located in the city's downtown area. The memorial's self-guided, interactive tour provides a detailed summary of the tragedy, and its outdoor reflecting pool and field of empty chairs underscore the tragedy's impact. The museum holds much more than heartrending artifacts such as broken clocks, chipped coffee mugs, and teddy bears. It also has a variety of official reports and studies of the bombing, including the City of Oklahoma's *Final Report: Alfred P. Murrah Federal Building Bombing April 19, 1995*. Prepared by the City Manager's Office, this four-hundred-page document presents a solid overview of the agencies and offices that responded to the bombing. Other materials at the museum include valuable information about the design and construction of the Alfred P. Murrah Federal Building and its eventual replacement—the Oklahoma City Federal Building. The University of Oklahoma in nearby Norman offers easy access to *The Daily Oklahoman*, the regional newspaper for the greater Oklahoma City area.

This work relies on oral history to supplement these sources. In the case of the Texas

City disaster, it draws largely from the recollections of survivors in *We Were There: A Collection of the Personal Stories of Survivors of the 1947 Ship Explosions in Texas City* (Texas City: Mainland Museum of Texas City, 1997), available at the Moore Memorial Public Library. This compilation of personal stories sheds light on the question of how working people along the waterfront reacted to the disaster and how the disaster affected their lives. Recollections of hospital workers who survived the San Fernando earthquake can be found in Shirley Braverman and Nancy Jenks, "California Quake," *American Journal of Nursing* 71 (April 1971): 708–12. Several of the MGM's former employees spoke to me about the 1980 fire. Most remained at the resort after Bally's Hotel and Casino purchased it in 1985, and the rest had retired or taken jobs elsewhere. A few of the latter employees were working at the Las Vegas Culinary Union—Local 226 of the Hotel Employees and Restaurant Employees International Union (HERE). The Oklahoma City's National Memorial and Museum helped me contact survivors of the 1995 bombing and offered the perfect place to interview them. The museum has a good oral history collection, much of it compiled by Marsha Kight, who lost her daughter in the tragedy. Kight's interviews form the heart of her book *Forever Changed: Remembering Oklahoma City, April 19, 1995* (Amherst, NY: Prometheus, 1998).

To locate congressional hearings relevant to disasters, researchers should consult the *US Congressional Committee Hearing Index* (Washington, DC: Congressional Information Service, 1981). Valuable electronic resources and databases were available through my home institution, the University of Hawai'i at Mānoa. Fortunately, the university subscribes to HeinOnline, a research-oriented product that provides access to a cornucopia of congressional records in a fully searchable format. Public hearings held after the 1956 air disaster and the 1971 San Fernando earthquake proved especially informative.

Index

Adams, Charles Francis, 23–34, 214n17
advocacy coalitions, 10, 12, 95, 128–29, 202
Aiken, George, 69, 221n63
airline industry: growth of, 75, 78–79, 83–85,
    198; in Los Angeles, 76, 79, 197; and plane
    cockpits, 79–80, 82–83, 94, 100; support of
    regulations, 94–95
Air Line Pilots Association (ALPA), 83–84, 92,
    94, 95, 103
air management system: and Grand Canyon
    disaster, 86–87; shortcomings of, 80,
    82–83, 93–96; transformation of, 76, 101–2
air traffic controllers: before 1956, 81–82,
    94–95; and Grand Canyon tragedy, 86–89;
    and new technologies, 96–97; post-tragedy
    reforms, 103
Air Transport Association (ATA), 94, 204
Airways Modernization Act (1957), 99
Airways Modernization Board (AMB), 97–99,
    101
Aldrich, Mark, 22
Alfred P. Murrah Building: background and
    design of, 166–71; federal workers within,
    173–74; operation and maintenance of,
    171–72; as poorly protected, 172–73, 193,
    202; private businesses in, 174–75; replace-
    ment of, 187–89
Allen, Clarence, 126, 203
Alquist, Alfred, 127–28, 130, 202
American Airlines, 78, 87
American Association of Blood Banks, 68
American Federation of Government Em-
    ployees (AFGE), 173
American Red Cross: and Long Beach earth-
    quake, 110; MGM Grand fire, 148, 153;
    Texas City disaster, 61
American Southwest. See Sunbelt West
America's Kids Child Development Center,
    166, 169, 174–75, 177, 179, 239n26
ammonium nitrate: disaster in Germany, 205;
    in Oklahoma City, 165, 198; subject to new
    regulations, 68, 206; in Texas City, 43–45,
    57, 64, 67, 72, 196, 198

Andrews, William, 96–97
Anglo-Norman tragedy (1850), 20
An Interesting Story (film), 14
Annan, Kofi, 195, 207
architects (and architecture): and fire safety,
    158; Murrah Building, 169–72; Oklahoma
    City Federal Building, 188–91; policy bro-
    kers, 203; seismic safety, 110–11, 126, 206
Arnold, Milton, 94–95
Ashcroft, John, 189–90
Ashtabula derailment (1876), 24–25
Aviation Facilities Planning Report, 97–98

Bakke, Oscar, 92
Balderston, Vern, 158, 204
Barney, Carol Ross, 188, 203
Beck, Ulrich, 5, 6, 16, 198, 199
Behncke, David, 83
Bellmon, Henry, 170
Bennett, Wallace, 101
Benninger, Fred, 150–51
Berman, Marshall, 14–15, 135, 208, 213n39,
    243n25
Birkland, Thomas A., 10
Blanck, Max, 34
Bliss, Philip, 24
Bolt, Bruce, 131
Brandt, Harvey, 126
Brown, Harold, 126, 229n63
Bush, George H. W., 163
Bush, George W., 190
butadiene, 47–48

Cahill, Robbins, 160–61
California Institute of Technology (Caltech),
    113, 126, 131
California Seismic Safety Commission, 130
Carnegie, Andrew, 31–32
Carnegie Relief Fund, 31–32, 216n38
Carter, Jimmy, 133
chemists: at Monsanto plant, 52–53, 64; and
    synthetic rubber, 47–48; and workplace
    disasters, 199

Cherry Mine disaster (1909), 28–29
Chicago, Illinois, 32, 37, 38, 67, 74, 78, 86, 188
Civil Aeronautics Act (1938), 80, 225n72
Civil Aeronautics Administration (CAA): early history of, 80–81, 83–84; and Grand Canyon tragedy, 87–89; and military, 80, 101–2; as poorly funded and outdated, 94; response to criticism, 83–84, 92, 96; subsumed by FAA, 97–99, 101, 225n72
Civil Aeronautics Board (CAB): establishment of, 80; and Grand Canyon tragedy, 92–93, 96–97; midair collisions of 1958, 100; rise of FAA, 101, 225n72
Cleveland Clinic fire (1929), 38–39
Clinton, Bill, 187, 202, 206
Coal Act (1969), 216n31
Collings, John, 90
Connally, John, 62
court rulings (and lawsuits): employer liability laws, 16, 18–19, 31, 41; Las Vegas resort fires, 157–58, 163, 235n56; and mining, 29; national regulative framework, 16, 41; New London school tragedy, 40–41; and steelmaking, 31–33; Texas City disaster, 70–71, 204, 221n65
Cranston, Alan, 132–33
Curtis, Edward P., 85, 97–98

Dalehite case, 221n65
Darwin, Charles, 13
Degenkolb, Henry, 131
Disaster Relief Act (1950), 70
disasters (general): definitions and life cycles of, 9, 196; as life-changing events, 186–87, 193–94; and media, 10, 12, 203–4; natural versus man-made, 104–5; as preludes to progress, 2, 12–13, 35, 41, 134, 163, 195, 199, 206–7; and service sector, 4, 17, 35–41, 195–96, 201; and social sciences, 9–11, 19, 102, 198; as unifying force, 179, 182, 186. See also workplace disasters
doctors (and surgeons): at Cleveland Clinic, 38–39; and Oklahoma City bombing, 177–79; after San Fernando Valley earthquake (1971), 105, 116–17, 121, 124, 200; and Texas City disaster, 61–62, 66, 71–72
Donnelly, Ignatius, 13

Earthquake Commission (Supervisor Hahn's), 125–27

Earthquake Council (Governor Reagan's), 130
Earthquake Hazards Reduction Act (1977), 132–33
earthquakes: Alaska (1964), 111–12; Anza-Borrego State Park (1968), 112, 127; Kern County (1952), 113; Lisbon (1755), 104–5; Long Beach (1933), 109–11; San Fernando Valley (1994), 133–34, 207, 232nn94–95; San Francisco (1906), 109; Santa Barbara (1925 and 1941), 109, 111. See also San Fernando Valley earthquake
Eastman, Crystal, 30–31, 43, 216n35
Eckert, H. K., 49–50, 53
Edison, Thomas, 14
Edwards, Don, 131
Eisenhower, Dwight, 71, 85, 97–99, 101
engineers (and engineering): and air travel, 75, 79, 83, 95–96, 102; in Grover Shoe Factory, 33; at Monsanto plant, 49, 52, 53, 57, 64, 68–69, 72; and New London school, 41; rail safety, 22–24, 74; seismic safety, 105, 110–15, 126–29, 131–34; on steamboats, 18; and steel mills, 32–33. See also Structural Engineers Association of California; US Army Corps of Engineers
entertainers (and entertainment), 18, 24, 29, 37–38, 153–54, 162

FAA. See US Federal Aviation Agency
Federal Aviation Act (1958), 100–101, 225n72
Federal Employees Credit Union (FECU), 174, 185–86, 239n25
fertilizer-grade ammonium nitrate (FGAN). See ammonium nitrate
Field, C. Don, 110
Field Act (1933), 110–11
firefighters: at Cleveland Clinic, 39; and MGM Grand fire, 142–43, 145–49, 157, 161, 163, 204; Texas City disaster, 44, 61, 68, 73, 220n39; Triangle shirtwaist fire, 34; at Veterans Hospital, 118
focusing event: as analytical concept, 10, 67, 202–4, 222n4; Grand Canyon air disaster, 75–76, 102–3; San Fernando Valley earthquake (1971), 105, 134; Texas City tragedy, 67
Foglietta, Tom, 187–88
Ford, Gerald, 133
Frankenstein (novel and film), 14
Freud, Sigmund, 13
Fritz, Charles, 9, 182
Furnas, C. C., 96